Gaming Media and Social Effects

The scope of this book series is inter-disciplinary and it covers the technical aspect of gaming (software and hardware) and its social effects (sociological and psychological). This book series serves as a quick platform for publishing top-quality books on emerging or hot topics in gaming and its social effects. The series is also targeted at different levels of exposition, ranging from introductory tutorial to advanced research topics, depending on the objectives of the authors.

More information about this series at http://www.springer.com/series/11864

Yiyu Cai · Wouter van Joolingen ·
Koen Veermans
Editors

Virtual and Augmented Reality, Simulation and Serious Games for Education

Springer

Editors
Yiyu Cai
Nanyang Technological University
Singapore, Singapore

Wouter van Joolingen
Utrecht University
Utrecht, The Netherlands

Koen Veermans
University of Turku
Turku, Finland

ISSN 2197-9685 ISSN 2197-9693 (electronic)
Gaming Media and Social Effects
ISBN 978-981-16-1360-9 ISBN 978-981-16-1361-6 (eBook)
https://doi.org/10.1007/978-981-16-1361-6

This Springer imprint is published by the registered company Springer Nature Singapore Pte Ltd.
The registered company address is: 152 Beach Road, #21-01/04 Gateway East, Singapore 189721, Singapore

Contents

Introduction

Koen Veermans, Wouter van Joolingen, and Yiyu Cai

Abstract This book is the result from continuous efforts to promote virtual and augmented reality, simulations and serious games for education. The 14 chapters of the book present a multi-facet view on educational applications of virtual and augmented reality, simulations and serious games across different educational systems, a wide age range and covering different domains. They also vary in their focus, with some describing the development of a single learning environment or assessment of such learning environments with learners, while others put more emphasis on the role of the teacher, either in design or implementation. What they share is the notion that they, despite different emphases, acknowledge that all these factors are important. Together the chapters in the book provide insights that will benefit researchers, developers and educators and will contribute to pedagogically well-designed virtual and augmented reality, simulation and serious game learning environments becoming natural elements in classrooms.

Keywords Augmented and virtual reality · Simulations · Games · Inquiry based learning · Teacher

1 Background

This book results from the fourth Asia-Europe Symposium on Simulation and Serious Games (EASSSG). The EASSSG alternates between Europe and Asia with the first held at Nanyang Technological University, Singapore [4], the second at Windesheim University of Applied Sciences, The Netherland [5], the third [6] in

K. Veermans (✉)
University of Turku, Turku, Finland
e-mail: koevee@utu.fi

W. van Joolingen
Freudental Institute, Utrecht University, Utrecht, The Netherlands

Y. Cai
Nanyang Technological University, Singapore, Singapore

Y. Cai et al. (eds.), *Virtual and Augmented Reality, Simulation and Serious Games for Education*, Gaming Media and Social Effects,
https://doi.org/10.1007/978-981-16-1361-6_1

Beijing Normal University at Zuhai as part of the ACM SIGGRAPH International Conference on Virtual-Reality Continuum and Applications in Industry and the fourth at the University of Turku, Finland. The symposium is devoted to providing insights for designers, researchers and practitioners that aspire to enhance the uptake of virtual and augmented reality (VAR), simulation and serious games in education. In the current situation, these insights may help to realize the potential benefits of VAR, simulation and serious games in classrooms across the world.

2 About the Book

VAR, simulations and serious games hold great promise for education. They have the potential to be used in student-centered learning environments where students actively engage with learning material leading to more integrated knowledge in comparison with teacher-centered learning environments. This aligns well with policies that advocate a shift from teacher-centered education toward more learner-centered education (e.g., [14]), advocate games and playful learning (e.g., [21]) and promote scientific thinking and evidence-based reasoning (e.g., [9, 15]). It could also help to address the problem of declining interest and motivation toward science [19] in combination with an expected increase of high qualification STEM jobs [8]. Despite this potential and findings that show their effectiveness (e.g., [3, 7]), VAR, simulation and serious games have yet to become a natural ingredient in most classrooms. Their uptake may have gotten another setback after the OECD reported that exposure to student-centered inquiry-based learning was associated with lower science scores [16]. While most of the references to this report ignored the positive effects on students' epistemic beliefs and motivation for science-oriented future careers that were also part of the report, the results and their interpretation themselves have also been questioned [2, 11].

There are several reasons that contribute both to the idea that student-centered inquiry-based learning may not benefit learning and to the uptake of VAR, simulation and serious games. One is that inquiry-based instruction is still sometimes considered as single category, while Hmelo-Silver et al. [10] already convincingly argued that there is a clear difference between supported and unsupported inquiry learning environments, something that has been further elaborated and backed up with evidence in a meta-analysis [12] and based on OECD data [2, 11]. In the case of AR, simulation and serious games, this support or guidance should be carefully designed in through the technology or provided by surrounding pedagogy and/or the teacher [13], and this is unfortunately not always the case. Bray and Tangney [20] for instance note that technology is mostly used as an alternative way of delivering the same content, which in the context of serious games has been brought up as an explanation for the fact that in most game-based interventions effects are small [3]. In the presence of other positive effects on, for instance, motivation (an often used argument for serious games), these small effects would not be problematic, but unfortunately, game-based learning is often not more motivating than traditional instruction [22]. This led Rodríguez-Aflecht et al. [17] to the conclusion that serious games should not be used because they are "believed to be motivating" but because of proven learning (or motivational) outcomes.

Another reason relates to assessment of learning outcomes. If one does not view VAR, simulation and serious games as an alternative way of delivering the same content, assessments of outcomes that are in line with the different view should also be considered. The above-mentioned motivation, the students' epistemic beliefs and motivation for science-oriented future careers are some examples that, in the context of the life-long learning and the policies that were mentioned earlier, are as or maybe even more important outcomes than the traditional learning outcomes.

The issues discussed above can be conceptualized through the relations, or absence thereof, between different elements of learning environments. The elements in Fig. 1 highlight the three main sources of support for students mentioned in Lehtinen and Viiri [13] and some contextual factors. All elements in Fig. 1 can in principle both be viewed as potential facilitators and challenges for learning depending on the way they have been taken into account in the realization of the learning environment.

As such, it can also provide a framework for assessing and positioning learning environments. For example, the pedagogy in many serious games is limited to individual drill and practice exercises where the role of teachers is either not considered or at best seen as the fairly passive deliverer of the learning material. As a result, these environments only address aspects of the curriculum that can be achieved through drill and practice. Curriculum goals that are more difficult to achieve are rarely addressed. It also makes that the designers of these games are rarely looking for affordances of games other than the assumed motivational effect which, if not realized, is problematic as gamification is not neutral [18].

The chapters in this book all explicitly or implicitly acknowledge the importance of some or many of the elements in Fig. 1 and aim to create active and engaging

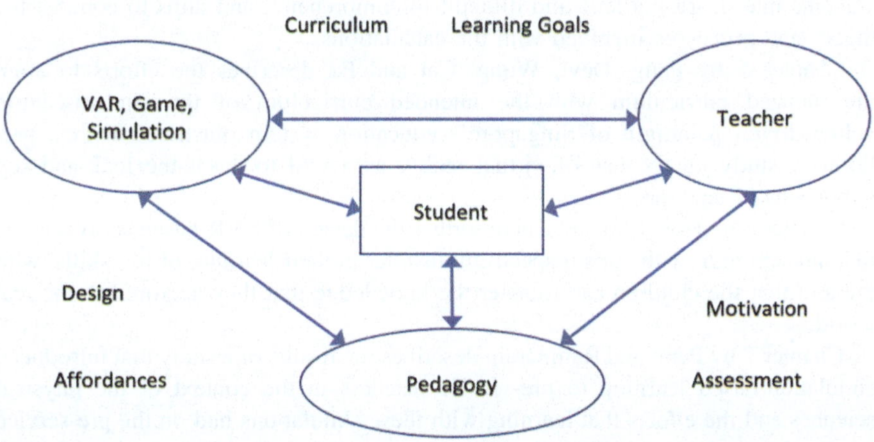

Fig. 1 Overview of contextual factors in VAR, simulation and serious games in education

learner-centered learning environments, but they differ in focus and emphases with regard to the way that elements in Fig. 1 are addressed. The chapter authors come from different parts of the world with different educational systems, learners range from elementary school students to pre-service teachers including special education students, and learning environments described in the chapters cover a wide variety of domains.

However, within this variety, many of the chapters also have a lot in common. The next section will first introduce the chapters and after that some of these commonalities and connections between the chapters on the basis of elements in Fig. 1.

3 Overview of the Chapters

This book has 14 chapters.

Chapter 2 by Brezovszky, Hannula-Sorminen, Veermans and Lehtinen describes the design of a math game that aims to develop adaptive number knowledge, an aspect that has proven difficult to develop in math education in classrooms. It addresses the cycle from design to large-scale evaluation in regular elementary school classrooms.

Chapter 3 by Papaevripidou, Hovardas and Zacharia describes an experimental study that investigated modeling of marine ecosystems comparing virtual and role-playing simulations and assessing their effects on tracking consequences of hypothetical changes in food webs and accounting for multi-directionality and uncertainty.

Chapter 4 by Gunawan, Kwan, Yang and Cai describes the design of an AR environment and accompanying assessment in the field of chemistry that aims to engage students in self-regulated learning. It addresses distillation, a topic that students may learn but still find difficult to comprehend, and aims to connect the theory and processes involved with the calculations.

Chapter 5 by Pang, Devi, Wong, Cai and Ba describes the efforts to align the realized curriculum with the intended curriculum of the student-centric values-driven paradigm of Singapore's education system illustrated with a preliminary study on the use of virtual reality as a tool to teach the lock-and-key hypothesis of enzymes.

Chapter 6 by Hoe, Cao and Cai describes the design of a VR-based serious game that aids children with autism spectrum disorder in their learning of life skills, with the aim that the children can transfer the knowledge that they acquired to the real world.

Chapter 7 by Penn and Ramnarain describes the results of a study that introduced simulation-based learning to pre-service teachers in the context of the physical sciences and the effects that learning with these simulations had on the pre-service teachers' attitudes toward the physical sciences.

Chapter 8 by Papaevripidou, Xenofontos, Hovardas and Zacharia examines pre-service teachers' capability to design inquiry learning activity sequences after receiving proper training. Pre-service teachers developed their own learning environment aligned with the national curriculum during the course and at the exam they were asked to transform a lesson plan into a learning environment.

Chapter 9 by Lehtinen, Hähkiöniemi and Nieminen describes how pre-service teachers provide guidance for students through questioning and by both structuring and problematizing student learning with simulations in the context of science lessons in elementary and math lessons in secondary education.

Chapter 10 by Veermans and Jaakkola describes the experiences of teachers implementing a simulation-based inquiry learning environment in their classroom. The learning environment has been well studied in research settings, and the aim was to learn more about the transition from research to classroom.

Chapter 11 by van Joolingen, Goei, Matimba, Pedro Gomes and Ba describes the design of an AR application for biology in secondary education, and how lesson study as an approach influenced design decisions during the process and provides a way to ensure the connection between the AR application and the curriculum.

Chapter 12 by Ba, Huang, Taib, Chen, Aung and Cai describes the AR design cardiac circulation, a topic that is difficult for students from both high school and colleges due its complexity. Being able to look inside this system is potentially a good way for students to learn about blood circulation from the perspectives of both structures and functions.

Chapter 13 by Tan and Tan describes a pedagogical and technical analysis of the potential of V360 in education, positioning in relation to VR, simulations and real, and highlighting similarities and differences. It also provides an overview of current technologies.

Chapter 14 by Ren and Zhang describes efficient face reconstruction and real-time facial expression for VR interaction that allows to create personalized avatars with facial expression that can be used in real-time VR environments in order to personalize the experience.

The 14 chapters are ordered to provide structure for the reader. In Chaps. 2–8 the focus is on learners, starting from learners in compulsory education and ending with pre-service teachers with the focus on their learning. The chapters on pre-service teachers (Chaps. 7 and 8) connect with the next chapters that also include pre-service or in-service teachers, but in these chapters, the focus is on their role in the implementation (Chaps. 9 and 10) and on the design (Chap. 11). The focus of Chaps. 12–14 is on design and technology.

Another way to view the chapters is along the elements in Fig. 1. All chapters address *Design* either explicitly by describing the design and decisions made in the process or implicitly by describing the features of learning environment in relation to the *Learning Goals*. Another element from Fig. 1 that comes back in most chapters is the connection with the *Curriculum*. In many cases, the explicit starting point in these chapters is something in the curriculum that has proven difficult to achieve otherwise and describe learning environments that attempt to address the issue. As a result, most of these chapters also have specific *Learning Goals* and try to use or exploit the *Affordances* of learning environments in a way that these learning goals are met (e.g., Chaps. 2, 3, 6 and 12). *Assessment* explicitly returns in some of the chapters but is also implicitly present in many others. Chaps. 2 and 4 for instance both discuss assessment explicitly as something that needs to be addressed in order to assess learners properly. In Chap. 9, the focus is on formative assessment during learning in a way that guidance is provided, but learner centeredness is also maintained. One aspect of assessment that is also included in several chapters is *Motivation*. Both Chaps. 5 and 7 address motivational aspects of the learning environments in their evaluation, while in Chap. 10, many of the teachers refer to these aspects when they reflect on implementing a learning environment in their classroom. The latter is one way that the role of the teacher is addressed, other ways are: including teachers in the design (Chaps. 5 and 11) the role of the teacher in supporting learners (Chap. 9) and educating future teachers (Chaps. 7 and 8). The last Chaps. (13 and 14) show some directions for the future. Together the chapters in the book provide insights on all elements in Fig. 1 that will hopefully benefit researchers, developers and educators and contribute to *Pedagogically* well-designed VAR, simulation and serious games becoming natural elements in classrooms.

References

1. All, A., Nuñez Castellar, E.P., Van Looy, J.: Assessing the effectiveness of digital game-based learning: best practices. Comput. Educ. **92–93**, 90–103 (2016). https://doi.org/10.1016/j.compedu.2015.10.007
2. Aditomo, A., Klieme, E.: Forms of inquiry-based science instruction and their relations with learning outcomes: evidence from high and low-performing education systems. Int. J. Sci. Educ. **42**(4), 504–525 (2020). https://doi.org/10.1080/09500693.2020.1716093
3. Boyle, E.A., Hainey, T., Connolly, T.M., Gray, G., Earp, J., Ott, M., Pereira, J.: An update to the systematic literature review of empirical evidence of the impacts and outcomes of computer games and serious games. Comput. Educ. **94**, 178–192 (2016). https://doi.org/10.1016/j.compedu.2015.11.003
4. Cai, Y., Goei, S.L. (eds.): Simulation, Serious Games and Their Applications. Springer, Singapore (2014)
5. Cai, Y., Goei, S.L., Trooster, W. (eds.): Simulation, Serious Games in Education. Springer, Singapore (2017)
6. Cai, Y., van Joolingen, W.R., Walker, Z. (eds.): VR, Simulations and Serious Games for Education, Springer Singapore (2019)
7. D'Angelo, C., Rutstein, D., Harris, C., Bernard, R., Borokhovski, E., Haertel, G.: Simulations for STEM learning: systematic review and meta-analysis. SRI International, Menlo Park, CA (2014)
8. European Commision: Does the EU need more STEM graduates? Final report. Publications Office of the European Union, Luxembourg (2015a). https://doi.org/10.2766/000444
9. European Commission: Science education for responsible citizenship (Report to the European Commission of the Expert Group on Science Education). European Union, Luxembourg, Europe (2015b). https://doi.org/10.2777/12626

10. Hmelo-Silver, C., Duncan, R., Chinn, C.: Scaffolding and achievement in problem-based and inquiry learning: a response to Kirschner, Sweller, and Clark (2006). Educ. Psychol. **42**(2), 99–107 (2007). https://doi.org/10.1080/00461520701263368
11. Kang, J.: Interrelationship between inquiry-based learning and instructional quality in predicting science literacy. Res. Sci. Educ. (2020). https://doi.org/10.1007/s11165-020-09946-6
12. Lazonder, A.W., Harmsen, R.: Meta-analysis of inquiry-based learning: effects of guidance. Rev. Educ. Res. **86**(3), 681–718 (2016)
13. Lehtinen, A., Viiri, J.: Guidance provided by teacher and simulation for inquiry-based learning: A case study. J. Sci. Educ. Technol. **26**(2), 193–206 (2017)
14. MOE: The Singapore Teaching Practice. https://www.moe.gov.sg/about/singapore-teaching-practice (2017). Accessed 9 Oct 2020
15. National Research Council. Next Generation Science Standards: For States, By States. Washington, DC: The National Academies Press (2013). https://doi.org/10.17226/18290.
16. OECD: PISA 2015 Results (Volume II): Policies and Practices for Successful Schools. OECD Publishing, Paris (2016)
17. Rodríguez-Aflecht, G., Jaakkola, T., Pongsakdi, N., Hannula-Sormunen, M., Brezovszky, B., Lehtinen, E.: The development of situational interest during a digital mathematics game. J. Comput. Assist. Learn. **34**, 259–268 (2018). https://doi.org/10.1111/jcal.12239
18. Toda, A.M., Valle, P.H.D., Isotani, S.: The dark side of gamification: an overview of negative effects of gamification in education. In: Cristea, A., Bittencourt, I., Lima, F. (eds.) Higher Education for All. From Challenges to Novel Technology-Enhanced Solutions. HEFA 2017. Communications in Computer and Information Science, vol. 832. Springer, Cham (2018)
19. Vedder-Weiss, D., Fortus, D.: Adolescents' declining motivation to learn science: a follow-up study. J. Res. Sci. Teach. **49**(9), 1057–1095. Simulations for science learning and dynamic geometry software for mathematics learning have opened up new possibilities for teaching and learning (2012)
20. Bray, A., Tangney, B.: Technology usage in mathematics education research—A systematic review of recent trends. Computers and Education, 114, 255–273(2017). https://doi.org/10.1016/j.compedu.2017.07.004
21. OECD: "Play!", Trends Shaping Education Spotlights, No. 18, OECD Publishing, Paris (2019). https://doi.org/10.1787/a4115284-en
22. Wouters, P., van Nimwegen, C., van Oostendorp, H., van der Spek, E. D.: A meta-analysis of the cognitive and motivational effects of serious games. Journal ofEducational Psychology, 105(2), 249–265 (2013). https://doi.org/10.1037/a0031311

The Number Navigation Game: An Overview of an Iterative Development Process

Boglárka Brezovszky, Koen Veermans, Minna Hannula-Sormunen, and Erno Lehtinen

Abstract The present chapter provides an overview of the iterative development and testing of the Number Navigation Game (NNG), a game-based learning environment which aims to strengthen primary school students' flexibility and adaptivity with arithmetic problem solving. The set of studies described in this chapter asked different questions using different methodologies through the phases of game development. However, the connecting principle was placing the integration of game features and mathematical content in the center of the process and developing valid and reliable means to operationalize the expected mathematical learning outcomes. Each step in the cycle served as a source of information to be used both in the development of the NNG and in the development of different types of measures of the mathematical learning outcomes. Conclusions of these steps are discussed in relation to the iterative game design process, the importance of using an integrated game design, and questions of measurement in the broader context of educational game design.

Keywords Game-based learning · Iterative game design · Arithmetic flexibility · Evaluation

---

B. Brezovszky (✉) · K. Veermans · M. Hannula-Sormunen · E. Lehtinen
Department of Teacher Education, University of Turku, Turku, Finland
e-mail: bogbre@utu.fi

K. Veermans
e-mail: koevee@utu.fi

M. Hannula-Sormunen
e-mail: mimarha@utu.fi

E. Lehtinen
e-mail: ernoleh@utu.fi

Y. Cai et al. (eds.), *Virtual and Augmented Reality, Simulation and Serious Games for Education*, Gaming Media and Social Effects,
https://doi.org/10.1007/978-981-16-1361-6_2

1 Introduction

After the initial excitement regarding the potential of using games for educational purposes, it is now clear that games as such do not represent a magic bullet. While most reviews find that game-based learning can be associated with higher learning gains when compared to more traditional methods of instruction, the difference is usually small [6, 19, 40, 41]. The explanation regarding this discrepancy between the potential of game-based learning and the lack of strong empirical results that would substantiate these claims is usually twofold.

Firstly, despite the affordances of technology to provide new ways of interaction with the educational content, most technology-supported educational tools are only used as side activities connected to traditional methods of content delivery [7]. In the specific domain of game-based learning, this means that the learning content is often not delivered by the gameplay, but the gameplay is used as an incentive to make students deal with the intended educational content [16–18]. This can create frustration and disappointment and can explain why game-based learning is often not found to be more motivating for students than traditional instructional methods [40] and why most game-based interventions are only marginally effective [6, 39, 40].

Second, there is often a disconnect between the type of skills and knowledge practiced within the context of a game and the means of measuring the expected learning outcomes [2, 3]. Measurement problems, such as using standardized tests to measure the outcomes of complex game-based training effects, relying solely on game performance measures, the overuse of quasi-experimental designs, and the lack of large-scale randomized studies are often raised as problems which hinder the estimation of the real educational potential of games [1, 3, 14, 15, 40, 41].

In order to make use of the potential affordance of games for educational purposes, it is important to both carefully design the integration of learning content and game features and have appropriate measures which can capture different aspects of the intended learning outcomes. Achieving this balance requires an iterative process of design and testing with the interaction of game design and educational content in the center [2, 35].

The current chapter aims to describe the process of development and cycles of testing of the Number Navigation Game (NNG), a game-based learning environment which aims to promote primary school students' flexibility and adaptivity with arithmetic problem solving. This chapter summarizes the results of a set of studies which used different methodologies to answer questions at different phases of development of the NNG. As a common theme, each study explored the relationship of game features and educational content within the game and how the learning gains associated with gameplay can be operationalized and measured both in the game and outside the game. Furthermore, the chapter provides a reflection on the use of iterations in developing game-based learning environments and on questions related to the integration of educational content and gameplay, measurement of learning outcomes, and the application game-based learning in the everyday classroom practice.

2 Background and Aims

In this chapter, the argument is made that the value of a game-based learning environment is strongest if there is a gap in teaching or learning a specific content for which the game-based format provides opportunities that were either difficult or not possible to achieve otherwise (e.g., [7]). In case of the NNG, this learning outcome was *flexibility and adaptivity with arithmetic problem solving*, which entails a strong mental representation and understanding of numerical relations and the efficient use of these relations when selecting problem-solving methods across various arithmetic contexts [4, 20, 23, 26, 37]. More specifically, the main aim of NNG practice was to strengthen students' *adaptive number knowledge*, which is a component of adaptivity with arithmetic and is described by the ability to recognize opportunities to use numerical relations during arithmetic problem solving [24, 25]. This type of mathematical expertise requires metal representation characterized by rich connections of numbers and arithmetic operations which promote the recognition of useful and efficient arithmetic strategies.

As studies show, direct instructional methods are insufficient in developing flexibility and adaptivity with arithmetic [12, 21] as the definition of what "useful" or "efficient" means can depend largely on the context, problem type, or individual preferences [20, 34, 37]. Therefore, developing methods to enhance flexibility and adaptivity with arithmetic should inherently promote an open-ended explorative type of mathematical practice. In this context, the transformative value [7] of the game-based format lies in the open-ended practice opportunities. In alignment with core aspects of flexibility and adaptivity with arithmetic [4, 23, 33, 34, 37], the basic structure of the NNG is open-ended, the game does not pre-define "optimal solutions," but it only provides a structure through its rules and design, which aims to promote reflection on the "usefulness" of various different solutions for the same problem. The main aim of the NNG game mechanics is to naturally enhance students' tendency to look for alternative solutions which is a prerequisite of developing a strong mental representation of numerical relations [28, 32].

In the regular mathematics instruction, this type of training might require individualized instruction and extra resources from teachers. The affordances of game-based learning, such as adaptivity, immediate feedback, and the scalability of practice, can provide new opportunities for developing these complex mathematical skills and knowledge. It was expected that intensive practice with the NNG, which requires working with different combinations of numbers and operations, solving arithmetic problems in multiple ways, and reflecting on solutions and their underlying arithmetical relations, would strengthen students' adaptive number knowledge and thus help to develop their flexibility and adaptivity with arithmetic problem solving.

Through the set of studies presented in this chapter, three main areas of game-based learning were explored: *game design, measurement of learning outcomes, and efficiency and application of the game-based training*.

The main aims of these studies were to

1. Operationalize adaptive number knowledge within a game-based learning environment and design a game for primary school students targeted primarily at developing adaptive number knowledge.
2. Operationalize adaptive number knowledge outside the game-based learning environment and show effects outside the game context.
3. Provide empirical evidence for the effects of the NNG training in developing primary school students' adaptive number knowledge and related mathematical skills first in controlled settings and later in the naturalistic classroom practice.

3 Methods

The choice for a game-based learning environment brings some inherent questions regarding design decisions, the integration of game features and learning content, measurement, and applicability of the training. To address these issues, the design was implemented as an iterative cycle of testing and development as it is presented on Fig. 1.

As Fig. 1 shows, in the piloting phase, an early prototype of the NNG was tested from a usability perspective focusing on the relationship of game features and the

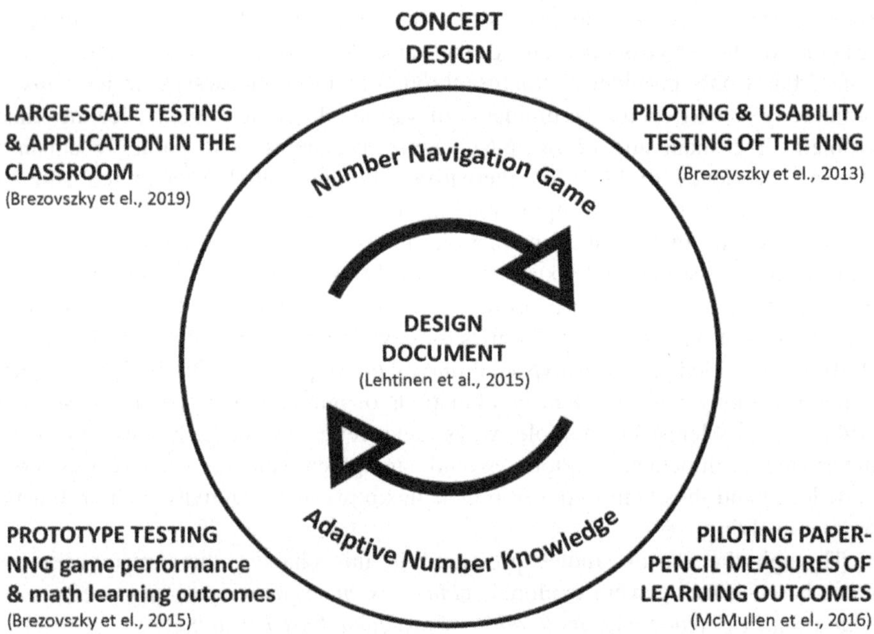

Fig. 1 Iterative development and testing process of the NNG [11], p. 22

mathematical learning content [8], and a series of pilot studies were conducted that aimed to develop and validate measures of adaptive number knowledge outside the game context [24]. The NNG was developed into a working prototype taking into account results from the initial play testing. This was followed by a small-scale pilot study using a second prototype of the NNG in which the study focused on the relationship of students' game performance and the development of their mathematical learning outcomes [9]. Findings from the piloting phase were used to develop both the NNG and in-game and paper–pencil measurements of the mathematical learning outcomes. The game design process and the relationship of the mathematical content and game features were documented and described in a design document [22], which served as a guideline in the following stages (see Fig. 1 in the center). In the final evaluation stage, a large-scale randomized control study was conducted using a final prototype of the NNG in three different grade levels [10].

4 Results

4.1 Relationship of Game Design and Mathematical Learning Content

To achieve complex learning outcomes using game-based learning, an *integrated game design* process is needed where the educational content is delivered through the core game mechanics [17, 29]. Accordingly, in the NNG, the aim was to integrate the open-ended, explorative mathematical practice within the core game structure so that each game feature triggers mathematically relevant actions, and the greater percent of students' interaction with the game equals to time on task with the intended learning content [17, 18, 29, 35].

The NNG was developed in the larger context of the studies presented in this chapter at the University of Turku Mathematical Learning Research Group. The game aims to improve primary school students' adaptive number knowledge by providing opportunities to engage in strategic work with various combinations of numbers and operations. The hundred-square (see Fig. 2) was selected as a basic external representation of natural numbers in the game, as it can support the development of number sense and numerical understanding [5, 31, 38]. Additionally, the ten by ten square structure was selected because it allows for both horizontal (number line) and vertical (base-10 system) representation and movement within those representations. Combining these two into diagonal movements can support decomposing and recombining numbers in various ways.

The game narrative uses the metaphor of navigation with a ship through different landscapes of land and sea where a map represents a basic game unit (Fig. 2). The players' goal is to collect different raw materials in order to build a settlement. Navigation is done by strategically selecting and executing different calculations to

Fig. 2 Screenshot of an NNG game map (harbor at number 14, target at number 57)

move the ship from one number to the next. An important aspect of the NNG design is that it makes reflection about alternative arithmetic solution methods a means of making progress within the game. The different game modes and rules aim to encourage players to use a large variety of number–operation combinations in their calculations, as well as look for different numerical relations and key numbers (i.e., numbers with several factors) within the natural number system.

In terms of game design, this type of game behavior is triggered, for example, by the different layout of each map where numbers covered by land cannot be used, by the different game modes which require players to adapt their strategies and use different restrictions in their calculations (e.g., reduce the number of moves between two numbers), or by random events like "pirate ships" which can appear to redirect players' routes and prompt them to use alternative number–operation combinations (for a more detailed description of the game features, see [22]). This variability makes reflection on the efficiency of each move (equation) a natural part of the gameplay, which is important because it is in line with the in situ, context-dependent definition of adaptivity and flexibility with arithmetic [33, 34, 37] and because reflection on the educational content is a much desired but hardly achieved part of educational game design [36].

When the NNG was tested in observational studies [8], results showed that the integrated game design was efficient in triggering the desired mathematical problem solving during gameplay. For example, even during a 1.5 h long play testing session, players executed around 200 mental calculations, and importantly, they also rejected even more alternative solutions while thinking about how to progress in the game. Players discussed and reflected frequently on possible solution options, and different game modes were efficient in prompting players to adapt their solutions strategies accordingly. These results suggest that in line with the idea of an integrated educational game design, both actual time on task and discussion around the game were mathematically relevant experiences.

4.2 Operationalization and Measurement of Learning Outcomes

In the initial stage, efficiency in the NNG was operationalized by the extent to which the game is able to enhance the type of mathematically relevant behavior in players [8]. This could include using a variety of numbers in their equations, using reflection and discussion when making moves, etc. [8]. Based on these results, it was assumed that the different NNG game features were efficiently triggering the expected mathematical behavior, and the amount of gameplay was considered as a measure of game performance. Thus, in future, quasi-experimental [9] and experimental studies [10] game performance was measured by the total number of maps completed in the NNG.

For better validity and reliability, it is suggested that learning outcomes related to game-based trainings are measured using a combination of different methods and types of measurements such as game-based and paper–pencil measures, near- and far-transfer tasks, and standardized tests [1, 3, 13, 27]. The main learning outcome associated with NNG is adaptive number knowledge, which was measured by the Arithmetic Production Task developed in the larger context of studies presented in this chapter [9, 10, 24, 25].

Like the NNG, the Arithmetic Production Task was also developed through several cycles of piloting (see [24]). These cycles of development were guided by the open-ended and in situ nature of flexibility and adaptivity with arithmetic. The main aim was to transfer this theoretical concept and parts of the practical training with the NNG in a paper–pencil format that can be flexibly applied over several grade levels and on a large scale. In the Arithmetic Production Task, students have to calculate a given target number by combining a set of 4–5 given numbers and the four arithmetic operations in as many ways as possible in a limited amount of time. Responses are coded on two criteria: (1) *quantity*, where the total number of correct solutions is counted; (2) *complexity*, where the total number of multi-step equations using different types of operations is counted. Alike the NNG, this format is open-ended and does not pre-define optimal solutions, but it provides a structure so that it is applicable and scalable in the classroom.

The relation of game performance and the paper–pencil measure of adaptive number knowledge was first piloted in a small-scale intervention study which showed a relation between the number of maps completed and students' complex solutions on the Arithmetic Production Task during post-test in grade six [9]. After this point, the Arithmetic Production Task was further refined with regard to the number and type of items used, and an application procedure was developed to ensure implementation fidelity on a large scale. In a large-scale randomized intervention study, results showed a relationship between game performance and both the amount and complexity of students' solutions on the Arithmetic Production Task [10]. Strengthening the results of the initial observational studies, the relationship found between game performance and mathematical learning outcomes highlights the importance of an integrated game design where time on task in the NNG equaled to mathematically relevant practice that could transfer to the outside game context as well.

4.3 Efficiency and Application of the NNG Training

The effects of the NNG training in developing flexibility and adaptivity with arithmetic problem solving were first tested in a smaller controlled quasi-experimental study [9] and then in a large-scale randomized intervention study in the naturalistic classroom context [10]. Both studies asked similar questions regarding the games' efficiency in developing adaptive number knowledge and related mathematical skills. As it was expected that students' adaptive number knowledge will be related to their general arithmetic fluency and pre-algebra knowledge [24, 25], these additional learning outcomes were measured by a combination of standardized and self-constructed tests. The Woodcock–Johnson math fluency sub-test was used to measure students' basic math fluency [30], and a measure of pre-algebra knowledge was used as a far-transfer measure of the relational thinking and the complex, multi-step mental work with arithmetic relations that can be foundational to algebraic thinking.

Results of both small- and large-scale intervention studies were promising and showed that training with the game was able to improve not only primary school students' adaptive number knowledge but also more basic arithmetic fluency and pre-algebra knowledge. Results of the large-scale intervention study [10] also revealed important grade-level differences in the development of the mathematical learning outcomes. The strongest results in developing adaptive number knowledge were observed in grade five. In grade four, practice with the NNG was more efficient in developing more basic arithmetic skills such as math fluency and the quantitative aspects of adaptive number knowledge. Finally, in grade six where more basic arithmetic skills are well established, results of the training could also be noticed on far-transfer tasks such as pre-algebra knowledge. These results suggest that the design of the NNG was flexible enough to support the development of different types of mathematical skills and knowledge across different grades, which are promising results with regard to the practical use and scalability of the training.

5 Conclusions

The aim of this chapter was to present the iterative design and testing process of the NNG game-based learning environment concentrating on questions related to game design, learning outcome measurement, and training application. Overall, the set of results described in this chapter show that it is possible to develop a game-based learning environment for classroom use that utilizes the affordances of gameplay to address aspects of learning (flexibility and adaptivity with arithmetic) that have proven difficult to address, otherwise, in the classroom. Secondly, results provide insights into important aspects of the design and evaluation of game-based learning environments that aided both the development of the NNG and the evaluation of its outcomes.

Overall, results strongly suggest that it is worth to invest resources into the integration of the educational content and the game mechanics, and it is important to consider integration as a general guiding principle in educational game design [3, 16, 17, 35, 41]. How well the content is integrated in the game, which features of the game will trigger the expected behavior or cognitive work, and how players interact with the relevant features are foundational questions as it is not the game itself but it is the design that makes game-based learning efficient or not [3, 7]. The design and development process should be iterative and multidisciplinary, and decisions should be based on theory and validated by empirical results.

The process of development and testing of the NNG described in this chapter is an example of such design and highlights a number of important aspects in this process. For instance, within these iterations, it is important to assess whether the design of the game triggers the intended activity. Additionally, it is important to validate whether this intended activity does indeed result in the intended learning outcomes. This may necessitate the development of new measurement instruments as was the case here (i.e., Arithmetic Production Task). However, the added value of developing new instruments is that the process can clarify the concept to be measured and how that concept can be broken down in terms of relevant behavior (i.e., using a variety of numbers and operation, discussion regarding the efficiency of the next move). In turn, this information can be used to guide the game design process in implementing game features which can trigger the expected behavior within the game (i.e., changing game modes, various landscapes, surprise elements, etc.). In the last phase, the controlled evaluation and large-scale classroom study served to first provide a proof of concept before providing evidence that the concept can also transfer to classroom practices.

While theoretically all the described steps in the cycle could be done during one evaluation, this type of iterative design was more efficient in case of the NNG, where in line with the theoretical propositions [4, 20, 34] both the training and the measurements of learning gains had to be flexible and open-ended. Naturally, the iteration process should not stop, but should continue during classroom use by utilizing the data of gameplay for refining the design.

References

1. All, A., Nunez Castellar, E.P., Van Looy, J.: Measuring effectiveness in digital game-based learning: a methodological review. Int. J. Serious Games, **1**(2), 3–21 (2014). https://doi.org/10.17083/ijsg.v1i2.18
2. All, A., Nuñez Castellar, E.P., Van Looy, J.: Towards a conceptual framework for assessing the effectiveness of digital game-based learning. Comput. Educ. **88**, 29–37 (2015). https://doi.org/10.1016/j.compedu.2015.04.012
3. All, A., Nuñez Castellar, E.P., Van Looy, J.: Assessing the effectiveness of digital game-based learning: Best practices. Comput. Educ. **92–93**, 90–103 (2016). https://doi.org/10.1016/j.compedu.2015.10.007
4. Baroody, A.J.: The development of adaptive expertise and flexibility: the integration of conceptual and procedural knowledge. In: The Development of Arithmetic Concepts and Skills: Constructive Adaptive Expertise ,pp. 1–33. Lawrence Erlbaum Associates Publishers, New Mexico (2003). https://doi.org/10.4324/9781410607218
5. Beishuizen, M.: Mental strategies and materials or models for addition and subtraction up to 100 in dutch second grades. J. Res. Math. Educ. **24**, 294–323 (1993). https://doi.org/10.2307/749464
6. Boyle, E.A., Hainey, T., Connolly, T.M., Gray, G., Earp, J., Ott, M., Pereira, J.: An update to the systematic literature review of empirical evidence of the impacts and outcomes of computer games and serious games. Comput. Educ. **94**, 178–192 (2016). https://doi.org/10.1016/j.compedu.2015.11.003
7. Bray, A., Tangney, B.: Technology usage in mathematics education research–a systematic review of recent trends. Comput. Educ. **114**, 255–273 (2017). https://doi.org/10.1016/j.compedu.2017.07.004
8. Brezovszky, B., Lehtinen, E., McMullen, J., Rodriguez, G., Veermans, K.: Training flexible and adaptive arithmetic problem solving skills through exploration with numbers: the development of number navigation game. In: 7th European Conference on Games Based Learning, ECGBL 2013, vol. 2, pp. 626–634 (2013)
9. Brezovszky, B., Rodríguez-Aflecht, G., McMullen, J., Veermans, K., Pongsakdi, N., Hannula-Sormunen, M.M., Lehtinen, E.: Developing adaptive number knowledge with the number navigation game-based learning environment. In: Describing and Studying Domain-Specific Serious Games, pp. 155–170. Springer International Publishing, Cham (2015). https://doi.org/10.1007/978-3-319-20276-1_10
10. Brezovszky, B., McMullen, J., Veermans, K., Hannula-Sormunen, M.M., Rodríguez-Aflecht, G., Pongsakdi, N., Lehtinen, E.: Effects of a mathematics game-based learning environment on primary school students' adaptive number knowledge. Comput. Educ. **128**, 63–74 (2019). https://doi.org/10.1016/j.compedu.2018.09.011
11. Brezovszky, B.: Using game-based learning to enhance adaptive number knowledge. Doctoral dissertation, University of Turku, Turku, Finland. Retrieved from https://www.utupub.fi/handle/10024/147025 (2019)
12. Canobi, K.H., Reeve, R.A., Pattison, P.E.: Patterns of knowledge in children's addition. Dev. Psychol. **39**(3), 521–534 (2003). https://doi.org/10.1037/0012-1649.39.3.521
13. Cheung, A.C.K., Slavin, R.E.: The effectiveness of educational technology applications for enhancing mathematics achievement in K-12 classrooms: a meta-analysis. Educ. Res. Rev. **9**, 88–113 (2013). https://doi.org/10.1016/j.edurev.2013.01.001
14. Connolly, T.M., Boyle, E., Hainey, T., Boyle, J.M.: A systematic literature review of empirical evidence on computer games and serious games. Comput. Educ. **59**(2), 661–686 (2010). https://doi.org/10.1016/j.compedu.2012.03.004
15. De Freitas, S., Oliver, M.: How can exploratory learning with games and simulations within the curriculum be most effectively evaluated? Comput. Educ. **46**(3), 249–264 (2006). https://doi.org/10.1016/j.compedu.2005.11.007

16. Devlin, K.: Mathematics Education for a New Era: Video Games as a Medium for Learning. A K Peters, Natick (2011)
17. Habgood, M.P.J.: The Effective Integration of Digital Games and Learning Content. PhD thesis, University of Nottingham, Nottingham (2007)
18. Habgood, M.P.J., Ainsworth, S.E.: Motivating children to learn effectively: exploring the value of intrinsic integration in educational games. J. Learn. Sci. **20**(2), 169–206 (2011). https://doi.org/10.1080/10508406.2010.508029
19. Hainey, T., Connolly, T.M., Boyle, E.A., Wilson, A., Razak, A.: A systematic literature review of games-based learning empirical evidence in primary education. Comput. Educ. **102**, 202–223 (2016). https://doi.org/10.1016/j.compedu.2016.09.001
20. Heinze, A., Star, J.R., Verschaffel, L.: Flexible and adaptive use of strategies and representations in mathematics education. ZDM-Int. J. Math. Edu. **41**, 535–540 (2009). https://doi.org/10.1007/s11858-009-0214-4
21. Hickendorff, M.: Dutch sixth graders' use of shortcut strategies in solving multidigit arithmetic problems. Eur. J. Psychol. Educ. **33**(4), 577–594 (2018). https://doi.org/10.1007/s10212-017-0357-6
22. Lehtinen, E., Brezovszky, B., Rodríguez-Aflech, G., Lehtinen, H., Hannula-Sormunen, M. M., McMullen, J., Jaakkola, T.: Number navigation game (NNG): design principles and game description. In: Describing and Studying Domain-Specific Serious Games, pp. 45–61. Springer International Publishing, Cham (2015)
23. Lehtinen, E., Hannula-Sormunen, M., McMullen, J., Gruber, H.: Cultivating mathematical skills: from drill-and-practice to deliberate practice. ZDM-Math. Edu. **49**(4), 625–636 (2017). https://doi.org/10.1007/s11858-017-0856-6
24. McMullen, J., Brezovszky, B., Rodríguez-Aflecht, G., Pongsakdi, N., Hannula-Sormunen, M. M., Lehtinen, E.: Adaptive number knowledge: exploring the foundations of adaptivity with whole-number arithmetic. Learn. Individ. Differ. **47**, 172–181 (2016)
25. McMullen, J., Brezovszky, B., Hannula-Sormunen, M.M., Veermans, K., Rodríguez-Aflecht, G., Pongsakdi, N., Lehtinen, E.: Adaptive number knowledge and its relation to arithmetic and pre-algebra knowledge. Learn. Instr. **49** (2017). https://doi.org/10.1016/j.learninstruc.2017.02.001
26. Nunes, T., Dorneles, B.V., Lin, P.-J., Rathgeb-Schnierer, E.: Teaching and Learning About Whole Numbers in Primary School. Springer International Publishing, Cham (2016). https://doi.org/10.1007/978-3-319-45113-8
27. Oprins, E.., Korteling, J.: Transfer of training of an educational serious game: the effectiveness of the CASHIER TRAINER. In: Cai, Y., Goei, S. (eds.) Simulations, Serious Games and Their Applications. Gaming Media and Social Effects. Springer, Singapore (2014). https://doi.org/10.1007/978-981-4560-32-0_15
28. Rittle-Johnson, B., Star, J.R.: Does comparing solution methods facilitate conceptual and procedural knowledge? An experimental study on learning to solve equations. J. Educ. Psychol. **99**(3), 561–574 (2007). https://doi.org/10.1037/0022-0663.99.3.561
29. Salen, K., Zimmerman, E.: Rules of Play: Game Design Fundamentals. The MIT Press, Cambridge (2003)
30. Schrank, F.A., McGrew, K.S., Woodcock, R.W.: Technical abstract (Woodcock-Johnson III Assessment Service Bulletin) No. 2 (2001)
31. Siegler, R.S., Ramani, G.B.: Playing linear number board games-but not circular ones-improves low-income preschoolers' numerical understanding. J. Educ. Psychol. **101**, 545–560 (2009). https://doi.org/10.1037/a0014239
32. Star, J.R., Rittle-Johnson, B.: Flexibility in problem solving: the case of equation solving. Learn. Instr. **18**(6), 565–579 (2008). https://doi.org/10.1016/j.learninstruc.2007.09.018
33. Threlfall, J.: Flexible mental calculation. Educ. Stud. Math. **50**(1), 29–47 (2002). https://doi.org/10.1023/A:1020572803437
34. Threlfall, J.: Strategies and flexibility in mental calculation. ZDM Math. Edu. **41**(5), 541–555 (2009). https://doi.org/10.1007/s11858-009-0195-3

35. Vanden Abeele, V., De Schutter, B., Geurts, L., Desmet, S., Wauters, J., Husson, J., Geerts, D.: P-III: a player-centered, iterative, interdisciplinary and integrated framework for serious game design and development. In: De Wannemacker, S., Vandercruysse, S., Clarebout, G. (eds.) Serious Games: The Challenge. ITEC/CIP/T 2011. Communications in Computer and Information Science, vol. 280, pp. 82–86. Springer, Berlin, Heidelberg (2012). https://doi.org/10.1007/978-3-642-33814-4_14
36. Veermans, K., Jaakkola, T.: Pedagogy in educational simulations and games. In: Cai, Y., van Joolingen, W., Walker, Z. (eds) VR, Simulations and Serious Games for Education. Gaming Media and Social Effects. Springer, Singapore (2019). https://doi.org/10.1007/978-981-13-2844-2_2
37. Verschaffel, L., Luwel, K., Torbeyns, J., Van Dooren, W.: Conceptualizing, investigating, and enhancing adaptive expertise in elementary mathematics education. Eur. J. Psychol. Educ. **24** (3), 335–359 (2009). https://doi.org/10.1007/BF03174765
38. Wilson, K.A., Bedwell, W.L., Lazzara, E.H., Salas, E., Burke, C.S., Estock, J.L., Conkey, C.: Relationships between game attributes and learning outcomes. Simul. Gaming **40**(2), 217–266 (2009). https://doi.org/10.1177/1046878108321866
39. Wouters, P., Oostendorp, H.V.: A meta-analytic review of the role of instructional support in game-based learning. Comput. Educ. **60**(1), 412–425 (2013). https://doi.org/10.1016/j.compedu.2012.07.018
40. Wouters, P., van Nimwegen, C., van Oostendorp, H., van der Spek, E.D.: A meta-analysis of the cognitive and motivational effects of serious games. J. Educ. Psychol. **105**(2), 249–265 (2013). https://doi.org/10.1037/a0031311
41. Young, M. F., Slota, S., Cutter, B., R., Jalette, G., Mullin, G., Yukhymenko, M. (2012). Our princess is in another castle: a review of trends in serious gaming for education. Rev. Edu. Res. **82**(1), 61–89. https://doi.org/10.3102/0034654312436980

The Effects of a Role-Playing Game and a Software Simulation Tool on Elementary School Students' Understanding When Modelling Feeding Relations in Ecosystems

Marios Papaevripidou, Tasos Hovardas, and Zacharias Zacharia

Abstract We designed and implemented two learning activity sequences with a focus on modelling and simulating trophic relations in marine ecosystems. Both implementations involved fifth graders, in two different samples (n = 24 for the first and n = 23 for the second sequence). The sequences differed only in the method used for undertaking modelling and simulations, which was a role-playing game, in the first sequence, and the Stagecast Creator software, in the second sequence. The sequences started with an introductory video, and then, students were asked to make a model of the observed phenomena in the form of a drawing. After revisiting preselected segments of the video and elaborating on modelling aspects of trophic relations, student models were refined in subsequent cycles of model-based inquiry. Students used either the role-playing game (first sequence) or Stagecast Creator (second sequence) to revise and enrich their models. Assessment tasks were administered in a pre-post arrangement to address inter-contextual transfer of student skills. Our findings revealed that the use of Stagecast Creator was more beneficial for students in terms of tracking the consequences of hypothetical changes in food webs and accounting for multi-directionality and uncertainty of effects across food webs.

Keywords Simulation · Ecosystem · Role-playing game · Computer software · Causal explanation

M. Papaevripidou (✉) · T. Hovardas · Z. Zacharia
Research in Science and Technology Education Group, Department of Educational Sciences, University of Cyprus, 75 Kallipoleos St, 1678 Nicosia, Cyprus
e-mail: mpapa@ucy.ac.cy

Y. Cai et al. (eds.), *Virtual and Augmented Reality, Simulation and Serious Games for Education*, Gaming Media and Social Effects,
https://doi.org/10.1007/978-981-16-1361-6_3

1 Introduction

Previous research has often highlighted a series of difficulties in developing student reasoning about ecosystems, especially in terms of elaborating on trophic relations [4]. The system thinking component which is involved in the domain of ecosystems has proven especially challenging even for educators themselves [8]. Although this type of knowledge and skills are challenging to establish and sustain, their implementation is crucial for sustainability considerations and decision-making for wildlife and natural resource management (e.g. [2, 3]).

Game simulations (e.g. [6]) or computer simulations (e.g. [11]) have been often recommended as effective applications to foster the development of student reasoning about trophic relations. Simulations are frequently accompanied by scaffolds to support the development of causal explanations [1]. In this study, we have employed a model-based inquiry approach to develop reasoning skills of primary school students about trophic relations in ecosystems. Specifically, we have compared two different learning activity sequences that involved different modelling strategies. The first sequence included a role-playing game, while the second sequence included a computer programming software. Model-based inquiry involved the subsequent tasks of model construction, testing and revision, in a trajectory which aimed to improve the model structure and simulate as closely as possible the modelled phenomenon [5, 7]. Our overall aim was to assess the two learning activity sequences in terms of their effectiveness in supporting students to develop causal explanations for trophic relations in ecosystems and discuss the implications of this assessment for learning and instruction.

2 Methods

2.1 Participants

Participants were fifth-grade students in two classes of a mid-sized suburban school in Cyprus. They were engaged in building, testing, revising, and re-testing models in two different learning activity sequences. The first sequence involved a role-playing game proposed in the national curriculum (n = 24), and the second sequence included Stagecast Creator, an agent-based computer modelling tool (n = 23). The latter was chosen because it offers a software environment designed for young learners to create microworlds based on a graphically represented program language (see [9]). None of the students received any formal training concerning modelling prior to the study. Students who used Stagecast Creator had used the simulation again in the past but not in the frame of model-based inquiry.

2.2 Procedure

Learning activities sequences are presented in Table 1 and included 15 activities each. Ten out of the 15 activities were identical, while the other 5 concentrated on the role-playing game in the first sequence (condition 1), and Stagecast Creator, in the second sequence (condition 2).

Both sequences started with a video on marine ecosystems, where we chose the award-winning documentary by BBC titled "The Blue Planet" (https://watchdocumentaries.com/the-blue-planet/?link=5). Students were then engaged in creating, testing, and revising models. The video served as the basic reference material to return to for outlining and taking into account additional scientific knowledge to test and revise student initial models. Model-based inquiry involved first drawings of marine ecosystems and moved on to incorporate the role-playing game in the first learning activity sequence and the Stagecast Creator in the second. The last activity in each sequence included several assessment tasks. The learning activity sequences were implemented by the first author and took place at the

Table 1 Learning activity sequence for conditions 1 (role-playing game) and 2 (Stagecast Creator)

	Learning activities in condition 1	Learning activities in condition 2
1	Watch a video on marine ecosystems	Watch a video on marine ecosystems
2	Draw a marine ecosystem	Draw a marine ecosystem
3	Compare drawings in groups	Compare drawings in groups
4	Discus comparisons in class	Discus comparisons in class
5	Revisit video segments to outline the basic needs of species	Revisit video segments to outline the basic needs of species
6	Revise drawing to account for the basic needs of species	Revise drawing to account for the basic needs of species
7	Discuss alternative models in class	Discuss alternative models in class
8*	Design a role-playing game	Build a model in Stagecast Creator
9	Gather additional information on food webs from video segments	Gather additional information on food webs from video segments
10*	Revise the role-playing game to account for food webs	Revise the Stagecast Creator model to account for food webs
11*	Undertake the revised role-playing game	Simulate the revised model in Stagecast Creator
12*	Compare video segments with the game simulation	Compare video segments with the Stagecast Creator model simulation
13	Draw a food pyramid model	Draw a food pyramid model
14*	Revise the role-playing game to account for food pyramids	Revise the Stagecast Creator model to account for food pyramids
15	Make predictions in a novel context of inquiry (forest ecosystem)	Make predictions in a novel context of inquiry (forest ecosystem)

Learning activities that differ between conditions are marked with an asterisk (*)

science laboratory (equipped with computers) and the yard of the school. Each sequence lasted approximately for eight forty-minute lessons.

In condition 1 ($n = 24$), students employed the role-playing game to advance their model-based inquiry. Students prepared a scenario with distinct role description to undertake the role-playing game. Each role corresponded to a species population in a prototype marine ecosystem, and involved trophic relations for each species. To reduce complexity to a manageable level, there were overall nine species included in the scenario: phytoplankton (taken to be one species), kelp, herring, crab, mussel, urchin, sea otter, seagull, and shark. When enacting the role-playing game, students used pieces of rope hanging from their waists to integrate trophic relations and represent the corresponding energy transfer. The game had a form of a run-and-chase series of actions. Every time a predator managed to touch a prey they acquired a piece of rope from them, thus adding to their energy level. To assign species to trophic levels and outline trophic relations, students went through selected segments of the video.

In condition 2 ($n = 23$), students used Stagecast Creator. Stagecast Creator allows students to model a phenomenon by means of a collection of visual before-after rules and programming by demonstration [10]. Therefore, it does not necessitate any syntactic language. The user can first choose the create a rule button to demonstrate a role and then select the character to be programmed (see Fig. 1). A rule maker window opens with the character depicted in its current state. The user then proceeds to enact the desirable behaviour of the character, for example, by dragging the character to a new position. The programming perspective behind this series of user actions is that rule construction is converted to a script monitored and modelled by the software. This script is to be performed anytime the criteria of its original design are met. Figure 1 presents the creation of a rule in Stagecast Creator for a herring character.

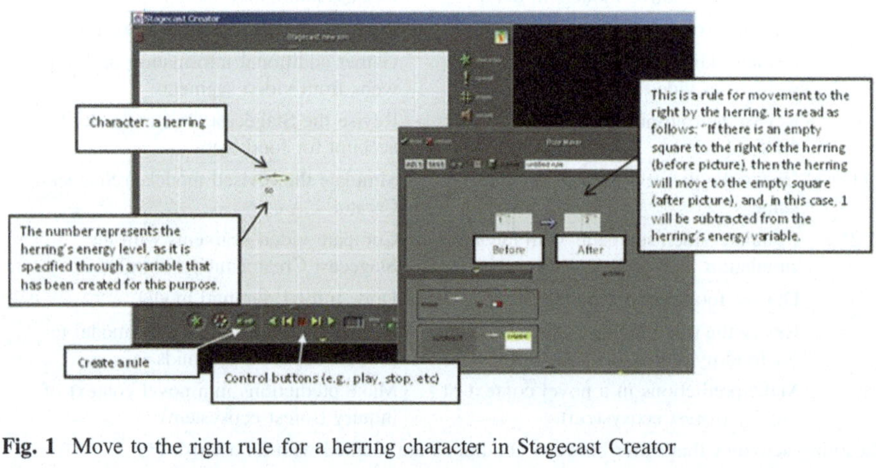

Fig. 1 Move to the right rule for a herring character in Stagecast Creator

2.3 Data Sources and Data Analysis

We developed nine tasks for assessing student reasoning, which was administered to students before and after instruction. Pre-test items also served to investigate any difference in prior knowledge and skills between condition, where there was not found any significant difference. Assessment tasks referred to forest ecosystems to present a transfer challenge for students, where they needed to implement knowledge and skills acquired in the instruction context (marine ecosystems) to a new context of inquiry (forest ecosystems). Assessment tasks were piloted and reviewed by an expert panel of six science educators.

Students first needed to construct a food web of a forest ecosystem by incorporating a set of pre-specified six species each only once (frog; snake; hawk; cat; sparrow; plant) and depict trophic relations by means of arrows starting from the prey species and ending to the predator species (assessment task 1). The rest of assessment tasks involved a readymade food web, which was used by students to elaborate on trophic relations depicted and on hypothetical developments introduced by a change in one population in the web. Assessment tasks 2–7 involved the depiction of trophic relations embedded in the food web for all different six species that were represented.

In assessment tasks 8 and 9, students were asked to predict the consequences within the forest ecosystem due to a hypothetical extinction of a second-class consumer (assessment task 8) and due to a hypothetical increase in the population of a first-class consumer (assessment task 9). To analyse student responses in these two tasks, we took into account patterns in student reasoning identified by previous research for hypothetical changes in trophic relations (e.g. [4]). These can involve non-scientific explanations (e.g. using aesthetic or anthropomorphic criteria to justify potential changes). Scientific accounts can be based on simple linear causality, where a cause is linked to an effect locally (e.g. a decrease of a producer population will cause the decrease of the population of a primary consumer species) or domino-like causality, where a cause is perceived to trigger a chain of subsequent changes within food webs (e.g. the decrease of a producer population will lead to a decrease of a population of a primary consumer species, and this will further cause the decrease of a secondary population species which feeds on the former). Uncertainty in the changes to be expected may also be added to causal explanations, for instance, when the hypothetical change in the food web in question will not necessarily eventuate along a certain direction only (e.g. a decrease of a producer may influence differently two different primary consumer species). The consideration of these additional and uncertain options reveals an advanced causal reasoning. The above categories of explanations guided our coding process. Two independent coders (first and last author) coded a subset of 10% of student responses in each assessment task. Inter-rater reliability (Cohen's k) amounted to over 0.89 across all assessment tasks, while disagreements between the two coders were resolved through discussion.

3 Results

A substantial majority of students in both conditions (22 out of 24 students in condition 1 and all 23 students in condition 2) were able, after instruction, to create a correct and complete food web in which all pre-specified species appeared only once, where the representation of all food relationships was correct and where the arrow that connected two species started from the prey and ended to the predator. Before instruction, no student was able to deliver a complete food web where each species had appeared only once. Similarly, almost all students in both conditions were able after instruction to identify and interpret the trophic relations that were embedded in a given forest food web (assessment tasks 2–7; students in condition 1 delivered 62.50% of trophic relations before instruction and 92.71% after; students in condition 2 delivered 60.31% of trophic relations before instruction and 96.74% after). Conditions did not perform differently across these items (1–7) after instruction. For each condition, improvement was highly significant across all items (Wilcoxon Signed Rank test *Zs* were all significant at $p < 0.001$).

Conditions differed in student ability to predict developments in the event of hypothetical population changes in the forest ecosystem food web (assessment tasks 8 and 9, see Tables 2 and 3, respectively). While the majority of students in condition 1 (role-playing game) could not progress beyond basic causal reasoning based on simple linear causality, the majority of students in condition 2 (Stagecast Creator) delivered after instruction an advanced causal reasoning based on domino-like causality with uncertainty. We need to highlight that all differences between conditions were highly significant (Mann Whitney Test *Z* significant at $p < 0.001$) but there was no significant difference between assessment tasks 8 and 9

Table 2 Student reasoning for predictions of hypothetical changes within an ecosystem for conditions 1 (role-playing game) and 2 (Stagecast Creator) for assessment task 8 (hypothetical extinction of a second-class consumer)

Type of reasoning	Category of response	Condition 1 before the course (n = 24)	Condition 1 after the course (n = 24)	Condition 2 before the course (n = 23)	Condition 2 after the course (n = 23)
Non-scientific reasoning	Aesthetic or anthropomorphic criteria	5	0	3	0
Basic causal reasoning	Simple linear causality	18	19	18	0
Elaborate causal reasoning	Domino-like causality	1	5	2	10
Advanced causal reasoning	Domino-like causality with uncertainty	0	0	0	13

Table 3 Student reasoning for predictions of hypothetical changes within an ecosystem for conditions 1 (role-playing game) and 2 (Stagecast Creator) for assessment task 9 (hypothetical increase in the population of a first-class consumer)

Type of reasoning	Category of response	Condition 1 before the course (n = 24)	Condition 1 after the course (n = 24)	Condition 2 before the course (n = 23)	Condition 2 after the course (n = 23)
Non-scientific reasoning	Aesthetic or anthropomorphic criteria	3	0	3	0
Basic causal reasoning	Simple linear causality	19	19	19	0
Elaborate causal reasoning	Domino-like causality	2	5	1	9
Advanced causal reasoning	Domino-like causality with uncertainty	0	0	0	14

within each condition, which implies that deviations in student performance in these cases needed to be attributed to the model-based inquiry strategy that was followed in each condition (i.e. role-playing game vs. computer software).

4 Discussion

A first major finding of our study was that both model-based inquiry strategies (role-playing game and computer software) proved equally successful in supporting students to undertake basic tasks in a transfer challenge, namely both learning activity sequences fostered student ability to create a food web with pre-specified species and to identify and interpret trophic relations included in a given forest food web in the novel context of inquiry (forest ecosystem). However, the curriculum based on Stagecast Creator as a modelling strategy led to improved causal reasoning for tropic relations compared to the role-playing game. This was reflected by the sharp divide between simple linear causality, which was exemplified by students in condition 1 (role-playing game) both before and after instruction, and domino-like causality to which students in condition 2 (computer software) had progressed after instruction. Indeed, the majority of students in condition 2 had also incorporated the component of uncertainty in their causal explanations, which is an exemplification of advanced causal reasoning.

Overall, the learning activity sequence with the Stagecast Creator modelling tool conveyed a deeper mastery of causal reasoning skills in contrast to the learning activity sequence with role-playing game (see also [9]). Given that the role-playing game is included in the national curriculum, there is an urgent need to revisit the curriculum and reconsider the proposed instruction for primary school teachers. Currently, there is a mismatch between curriculum standards, especially in terms of addressing causal reasoning for trophic relations in ecosystems, with the instructional directions provided. The addition of computer software will be an insightful idea for updating the learning activity sequences proposed. The better performance of students who used Stagecast Creator may be probably due to an increased ownership in model building, which may be linked to the programming functionalities and skills it necessitates compared to the role-playing game. These aspects need to be further investigated by future research. In addition, future research could assess the implementation of different combinations of the role-playing game and the computer software with the same group of students (see for instance [6]) and outline the added value of each in establishing student reasoning about trophic relations.

References

1. Basu, S., Sengupta, P., Biswas, G.: A scaffolding framework to support learning of emergent phenomena using multi-agent-based simulation environments. Res. Sci. Educ. **45**, 293–324 (2015). https://doi.org/10.1007/s11165-014-9424-z
2. Bravo-Torija, B., Jiménez-Aleixandre, M.-P.: Progression in complexity: contextualizing sustainable marine resources management in a 10th grade classroom. Res. Sci. Educ. **42**, 5–23 (2012). https://doi.org/10.1007/s11165-011-9254-1
3. Gresch, H., Bögeholz, S.: Identifying non-sustainable courses of action: a prerequisite for decision-making in education for sustainable development. Res. Sci. Educ. **43**, 733–754 (2013). https://doi.org/10.1007/s11165-012-9287-0
4. Grotzer, T.A., Bell-Basca, B.: How does grasping the underlying causal structures of ecosystems impact students' understanding? J. Biol. Educ **38**, 16–29 (2003). https://doi.org/10.1080/00219266.2003.9655891
5. Halloun, I.A: Modeling Theory in Science Education. Springer, The Netherlands (2006)
6. Hovardas, T.: A learning progression should address regression: insights from developing non-linear reasoning in ecology. J. Res. Sci. Teach. **53**, 1447–1470 (2016). https://doi.org/10.1002/tea.21330
7. Hovardas, T., Pedaste, M., Zacharia, Z., de Jong, T.: Model-based science inquiry in computer-supported learning environments: the case of Go-Lab. In: Azad, A.K.M., Auer, M., Edwards, A., de Jong, T. (eds.) Cyber-Physical Laboratories in Engineering and Science Education, pp. 241–268. Springer, Dordrecht (2018)
8. Lee, T.D., Jones, M.G., Chesnutt, K.: Teaching systems thinking in the context of the water cycle. Res. Sci. Educ. (2017). https://doi.org/10.1007/s11165-017-9613-7
9. Papaevripidou, M., Constantinou, C.P., Zacharia, Z.C.: Modeling complex marine ecosystems: an investigation of two teaching approaches with fifth graders. J. Comput. Assist. Learn. **23**, 145–157 (2007). https://doi.org/10.1111/j.1365-2729.2006.00217.x
10. Smith, D.C., Cypher, A., Tesler, L.: Novice programming comes of age. Commun. ACM **43**, 75–81 (2000)
11. Wilensky, U., Reisman, K.: Thinking like a wolf, a sheep, or a fire-fly: learning biology through constructing and testing computational theories—an embodied modeling approach. Cogn. Instr. **24**, 171–209 (2006). https://doi.org/10.1207/s1532690xci2402_1

Augmented Reality Application for Chemical Engineering Unit Operations

Poernomo Gunawan, James Kwan, Yiyu Cai, and Rui Yang

Abstract Understanding unit operations is an essential part in chemical engineering course. An important example is the continuous distillation column, which operation is often seen as a black box, where the incoming feed will undergo separation process inside the column to produce desired products. Despite having learned the concepts on how they work, students may find it difficult to comprehend and visualize what is going on inside a distillation column and how to connect various theories involved in the design and calculations. By developing a virtual visualization tool, such as augmented reality (AR), students can better visualize the process, such as fluid flow profiles and different components that make up a distillation column. Although the idea of incorporating AR for higher education learning is not entirely new, this is the first initiative to implement virtual technology for chemical engineering curriculum in Singapore, which serves as a novel pedagogical approach to complement the conventional pen-and-paper teaching method. Besides enhancing the students' learning experience, it is believed that the AR application would provide a way to improve the students' motivation and interest to learn the subject as well as a complementary tool for laboratory demonstration, as it is practically safe and time-saving.

Keywords Augmented reality · Unit operations · Chemical engineering

P. Gunawan (✉)
School of Chemical and Biomedical Engineering, Nanyang Technological University Singapore, 62 Nanyang Drive, Singapore 637459, Singapore
e-mail: pgunawan@ntu.edu.sg

J. Kwan
Department of Engineering Science, University of Oxford, Oxford, UK
e-mail: james.kwan@balliol.ox.ac.uk

Y. Cai · R. Yang
School of Mechanical and Aerospace Engineering, Nanyang Technological University Singapore, 50 Nanyang Avenue, Singapore 639798, Singapore
e-mail: myycai@ntu.edu.sg

Y. Cai et al. (eds.), *Virtual and Augmented Reality, Simulation and Serious Games for Education*, Gaming Media and Social Effects,
https://doi.org/10.1007/978-981-16-1361-6_4

1 Introduction

1.1 Augmented Reality Application for Education Purposes

With the advancement of information and communication technologies affecting more aspects of society, adoption of computer technologies such as virtual learning by higher educational institutions has been gaining momentum. It is believed to transform the future landscape of teaching and learning activities, as reported in several literatures [1–4]. In the rapidly changing environment, students not only need to acquire the competencies of creativity, innovation, collaboration, and critical thinking, but they also need real-world experiences to help them link theory and practice [4]. By bridging virtual and real worlds together, in particular, augmented reality (AR) allows the combination of real-world elements captured through a camera with multimedia elements such as text, images, video or 3D models, and animations [3]. The coexistence of virtual objects and real environments is believed to allow learners to visualize complex spatial relationships and abstract concepts [5], experience phenomena which otherwise impossible in the real world [6], interact with two- and three-dimensional synthetic objects in the mixed reality [7] and develop important practices and literacies that cannot be developed and enacted through other technology-enhanced learning environments [8]. Given all of these benefits, AR has been used in various fields, such as science [9, 10], medicines [11, 12], arts, and architecture [13], as an educational tool in both formal and informal settings. Research has shown that the merging of AR with education may potentially make learning more active, effective, and meaningful because it allows students to immerse in a more realistic experience [14, 15]. It is in line with ICAP framework where the students' learning improves, as they are more engaged with the learning materials [16].

Visualization is a key factor in facilitating understanding and preventing misconceptions in education. As a result, many find science and engineering difficult due to the inability to properly visualize the abundant abstract concepts, which lead to misconceptions in these fields. The use of AR can minimize and correct these misconceptions because it allows macro- or micro-visualization of objects, displaying objects, and concepts from different perspectives to enhance a student's understanding [17]. For such reason, there are several examples of virtual reality (VR)-based learning environments in the field of engineering education, such as mining engineering, construction engineering, manufacturing, and chemical engineering [18–20]. In particular, Ludwig-Maximillians University developed mixed reality application which augments the thermal flux in metal by overlaying a representation of temperature as false-color visualization directly onto the object [20]. University of Rochester developed AR to simulate chemical reactors configuration in a chemical plant by using coffee mugs to represent virtual 10-cubic meter reactors [21]. Their success inspired the School of Chemical and Biomedical Engineering at Nanyang Technological University to develop its pioneer virtual application to complement in-class teaching. Considering the ubiquitous access to smartphones and other mobile devices as well as wireless connection [22, 23], the aim is to leverage on these technologies for constructive utilization of mobile AR application during lecture and laboratory sessions.

With regards to the development of AR learning features, various instructional approaches have been taken into account in the design of AR learning environments, including place-based learning [24], role playing [25], task-oriented learning [8, 25], and problem-based learning [8, 26]. In the current approach, the latter is incorporated into AR by assigning randomized tasks for students to solve through which they can test their ideas and apply problem solving skills.

1.2 Augmented Reality Application in Chemical Engineering

This chapter will focus on the development and introduction of AR application in the curriculum of Chemical Engineering undergraduate course, in particular, unit operations, a core subject dealing with complex separation processes, such as fractional distillation, gas absorption, liquid–liquid extraction, etc. Continuous distillation is an essential purification system in the petrochemical industries and refineries that works principally based on the difference in the degree of volatility of the components present in the mixture. Conventionally, it has been taught in class as black box diagrams or 2D models comprising incoming feed stream and outgoing top and bottom product streams as shown in Fig. 1, which lack user visibility, thus preventing students from developing a deeper intuition of the systems and concepts at play. While laboratory sessions help to bridge the theory with practical application to a certain degree and provide hands-on experience, not every student has a chance to observe the internals of the column and to thoroughly operate the

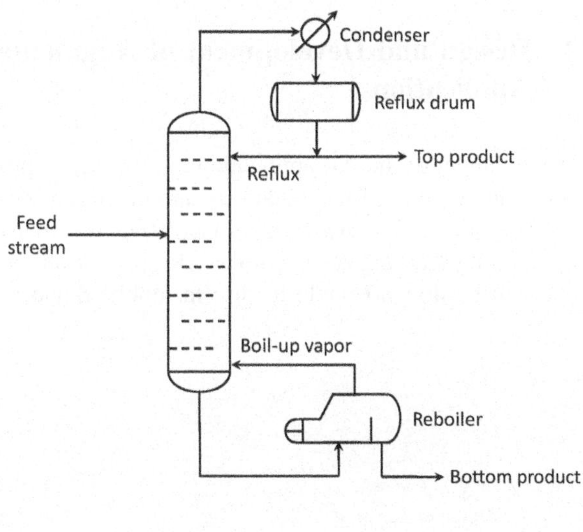

Fig. 1 Schematic diagram of a continuous distillation column

equipment due to limited number of the apparatus and the relatively short laboratory session. Therefore, despite having learned the fundamental concepts, students may still find it difficult to comprehend and to visualize what is going on inside a distillation column, such as the fluid flow and composition profiles, to connect theories involved in the column design and calculations, in particular mass transfer between different phases and liquid–vapor equilibrium, as well as to understand the correlations between various process parameters and the resulting products purity.

Given the limitations of the conventional teaching method, which primarily relies on the derivation of equations, plotting on graphs, and 2D illustration of the setup, it was therefore proposed to implement AR application as an initiative to incorporate virtual technology that will promote better visualization and learning of continuous distillation process. In addition, the AR application can also be used as a "dry" experimental tool for laboratory teaching, as it is practically safe, independent of physical and chemical constrain, and time-saving. Having a virtual experiment to go with the physical one will make the laboratory session more meaningful, as students can instantaneously compare and evaluate their observations with the simulation results, hence, promotes more in-depth discussions on the subject matter. This experiential learning is believed to improve students' motivation and engagement in learning the subject.

With the aforementioned motivations, the project aims to achieve the following outcome:

1. Build up the capability and asset in developing AR for Chemical Engineering curriculum at Nanyang Technological University.
2. Promote students' self-learning and self-evaluation of the prior concepts through the simulation and assessment scenes embedded in the AR application.
3. Evaluate the effectiveness of AR application in engaging students in their learning and improving their interest in unit operations.

2 Design and Development of Augmented Reality Application

Development of the AR application in this project primarily consists of four stages: (1) setting up the software platform and the triggering object, i.e., marker image that activates the AR experience, (2) creating the model of the distillation column, (3) creating learning scenes in the AR application, and (4) building the application that will be downloadable for use in mobile devices.

Fig. 2 **a** Distillation column setup as image target and **b** the augmented features on the target image

2.1 *Software Platform and the Triggering Object*

Unity version 2018.2.14f was being used as the primary software for developing the AR application in this project. Being a popular platform for 2D and 3D game development, Unity offers free personal version for students to download in their mobile devices, which is a key consideration for using the software. Vuforia was used as complementary software to support integration of advanced computer vision functionality for AR development on various devices with different systems such as Android, iOS, and windows. More importantly, it is a free asset on unity platform and offers several ways to set up an AR environment by using different kinds of triggering object, such as image target, model target, ground plane, etc. In this project, the image target and the physical distillation column setup in the School laboratory were used as the triggering object as shown in Fig. 2. A decent quality target image is necessary for the AR camera to detect and recognize the augmented features.

2.2 *Creating the Model for the Distillation Column*

The 3D model of distillation column was built on the unity platform, and it was based on the physical laboratory setup comprising the following components:

a. Feed tank containing liquid ethanol–water mixture with composition of 50 mol%. The feed stream was pre-heated to form saturated liquid prior to entering the column at a specific location in the middle of the distillation column.
b. Distillation column with a fixed number of eight sieve trays over which the distillation process occurs.
c. A condenser to condense the vapor leaving the top of the distillation column to liquid phase. The liquid leaving the condenser was collected in a reflux drum, from which a fraction of it was sent back to the distillation column as reflux, whereas the remaining was drawn as distillate (top product). It had higher ethanol composition than the feed.
d. A reboiler to partially vaporize the liquid leaving the bottom of the distillation column. The temperature of the reboiler was higher than that of the condenser. The resulting vapor was sent back to the column, while the hot liquid was drawn as bottom product, which had lower ethanol composition than the feed.

In this project, SOLIDWORKS software was used to create the model. It was further converted to the mesh model required in unity platform to form a 3D model as shown in Fig. 3.

Fig. 3 Distillation column model created on the Unity platform

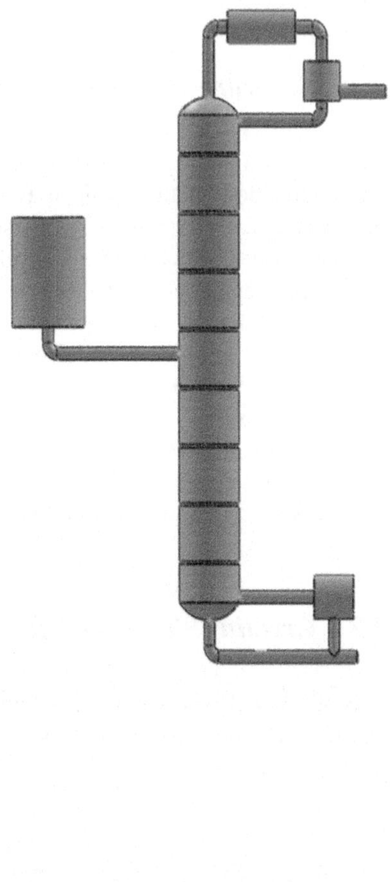

2.3 *Creation of the Learning Scenes*

Following the development of the model, three learning scenes were subsequently created to address different learning objectives:

a. Introductory scenes to introduce the concept of distillation process and different components that make a distillation setup;
b. Simulation scene to learn the effect of distillation parameters, mainly reflux ratio, boil-up ratio, and the feed entry location on the products composition;
c. Assessment scene to test student's ability to solve a problem that is randomly generated by the application for each individual student.

2.3.1 Introductory Scenes

The objective of the introductory scenes is to present and provide visualization on the working principles of a distillation column as well as to understand the role of each component that makes up a distillation unit. Figure 4 shows the first scene that appears upon activating the target image or physical equipment setup, which resembles a black box model. Upon pressing the "Start" button, students are able to disassemble the 3D model into four main sections, namely feed tank, distillation tray column, condenser, and reboiler, as presented in Fig. 5. The disassembled model was made transparent to give better visualization on the column internals and the movement of the liquid and vapor throughout the system. Intuitively, students need to press the "Flow" buttons corresponding to each respective component in the correct order to operate a "dry experiment." Otherwise, students will encounter a warning and unable to proceed. First, the empty column is to be filled with the ethanol solution from the feed tank, which is flowing down to the reboiler. The reboiler then heats up the liquid and partially vaporizes it. The resulting vapor is subsequently flowed to the bottom of the column where it will rise up and be in contact with the liquid on each tray. The vapor will then flow into the condenser and turned to liquid as depicted in Fig. 6. The condensed liquid is collected in a reflux drum, and a fraction of it was fed to the top of the column. By doing this activity, students are able to observe the changes in terms of ethanol composition and liquid–vapor flow patterns in the column, condenser, and reboiler through the color degradation from dark blue at the bottom of the column and reboiler that corresponds to low ethanol composition to light green at the top of the column and condenser that corresponds to high ethanol composition. This simulation thus offers good visualization as the physical liquid solution used in the experiment is colorless, making it difficult for students to monitor the composition change on each tray.

Figure 7 shows a detailed image of the interaction between liquid and vapor through a sieve tray. By applying color degradation, students could visually observe how water is partially condensed from the vapor to the liquid phase, whereas ethanol is partially vaporized from the liquid to the vapor phase, leading to the

Fig. 4 First scene upon activating the target image or the physical laboratory setup

ethanol-rich vapor leaving the tray. This feature enables students to better grasp the concept of the materials transfer between liquid and vapor phase that occurs on the tray.

2.3.2 Simulation Scene

The simulation scene allows students to independently explore the effect of three main parameters of distillation process, such as boil-up ratio, reflux ratio, and the feed entry location on the top and bottom products compositions. Referring to Fig. 8, the boil-up ratio (VB) is varied 2.5, 5.0, and 10, while the reflux ratio (R) is varied 0.1, 1.0, and 10. Theoretically, there is a correlation between the former and the latter, which students need to identify by observing their effects on the column performance. In addition, students also have the freedom to select their own feed entry location by shifting the feed tank as depicted in Fig. 8c, d. After selecting the parameters, students run the simulation by pressing "Game On" button. The simulation result at steady state is shown in Fig. 9, where the compositions along the column are displayed together with that of the top and bottom products. Students can repeat this activity as many times as necessary to foster their learning and understanding. By completing this section, students are expected to achieve the objective to find the following correlations:

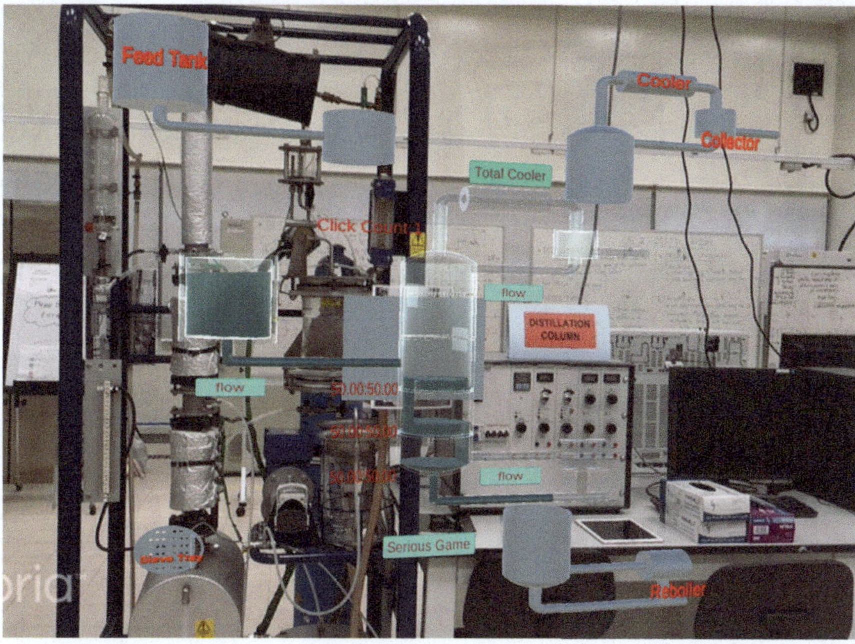

Fig. 5 Disassembled scene of the distillation setup

a. Reflux ratio, R is directly proportional to the top product purity;
b. Reboiler ratio, VB is inversely proportional to the top product purity;
c. Feed entry stage does not strongly affect the top product purity.

2.3.3 Assessment Scene

The assessment scene is a follow-up on the simulation scene. As shown in Fig. 10, the AR application generates a unique problem statement for each student who logs in to this section, such as to determine the correct distillation parameters to achieve a certain top product purity and flow rate. Students are able adjust the feed entry location and perform the calculation. To solve the problem, students are supposed to do hand calculation prior to entering the correct values of the required parameters. To facilitate self-evaluation on their understanding of the concept and calculations of distillation column, students are given multiple number of attempts, and they can see the score at the end of the session. More number of attempts could indicate that students may not understand the calculation procedures completely, and they may need to further review the concept. For monitoring purposes, the students' identifier and the number of attempts are recorded in a database. By accessing the database, the instructor can therefore assess the students' performance

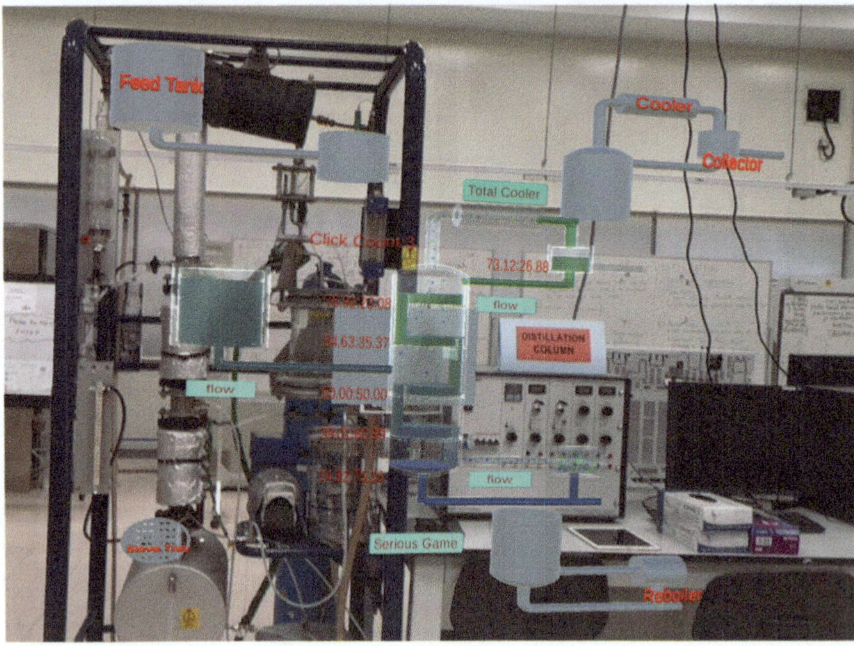

Fig. 6 Composition distribution throughout the distillation column is indicated by color change scheme

Fig. 7 Magnification of the distillation column visualizing the contact of liquid and vapor on each tray

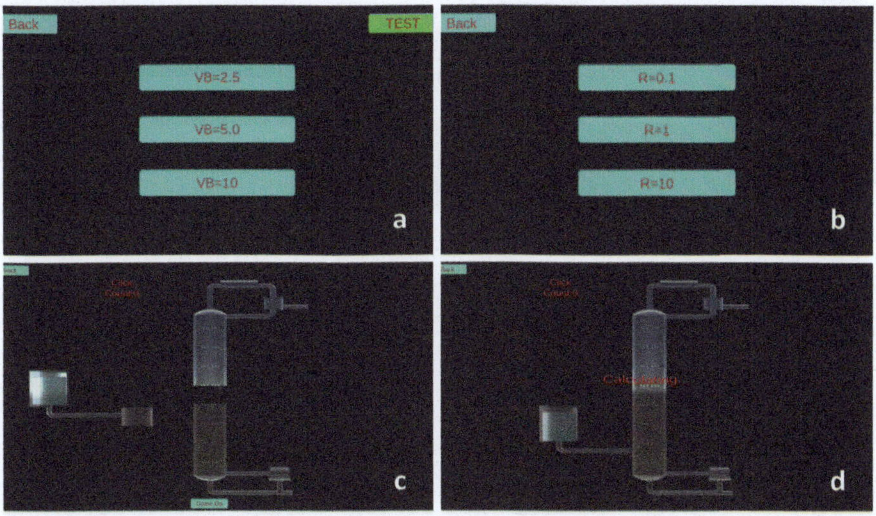

Fig. 8 Selection of distillation parameters: **a** boil-up ratio VB, **b** reflux ratio *R*, **c** feed entry location, and **d** running the simulation

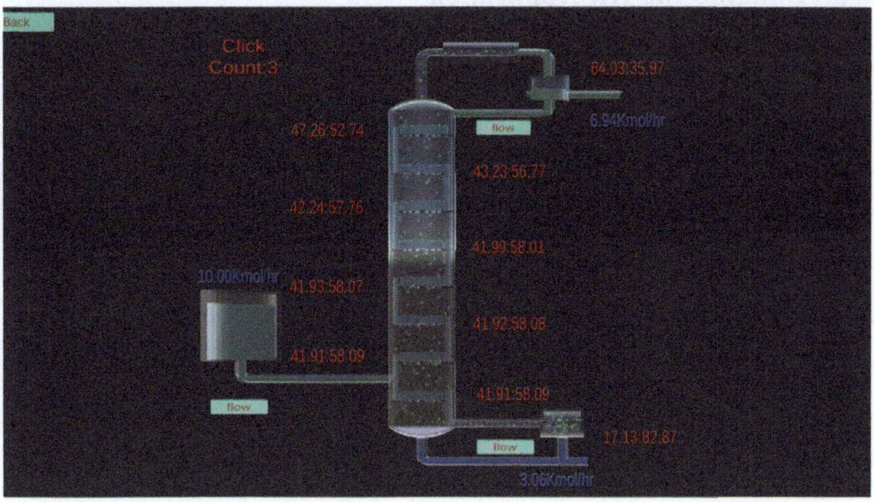

Fig. 9 Calculation results from the simulation scene

and render any necessary assistance to those who take more attempts to solve the problem. In addition, the instructor is also able to get insights on the effectiveness of the teaching in class.

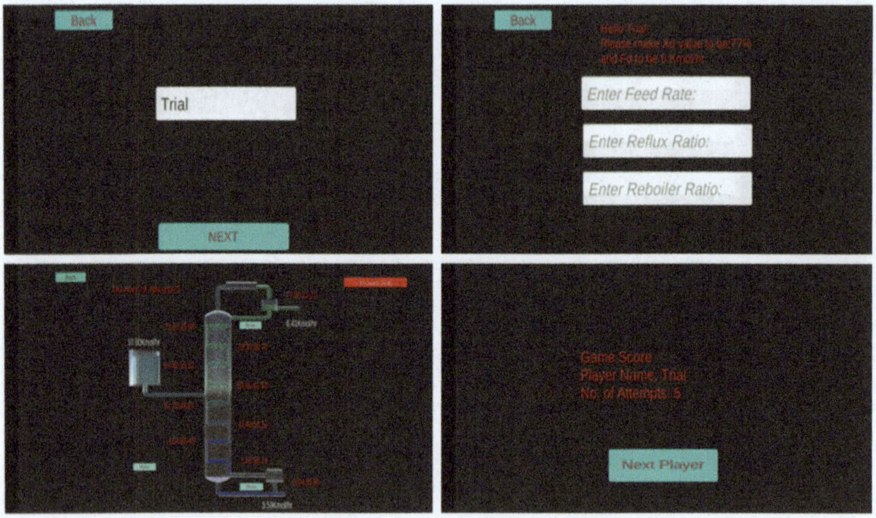

Fig. 10 Problem statement generated by AR application in the assessment scene

2.4 *Building the AR Application*

When the AR development was completed, all of the necessary scenes were selected. The unity project environment was then switched to Android and built the content in an APK file in order to install the application in an Android mobile device for the students to use. To ensure greater flexibility, the application setting was further extended for iOS users.

3 Implementation of AR Application and Students Feedback

The mobile AR application was first implemented in Aug−Oct 2019 to a class of over 100 students who were enrolled in Chemical Engineering Unit Operations. Since it was the first initiative within the School and the prototype only contains basic principles of distillation, it was primarily introduced as a complementary tool for the in-class lectures where students explored the application at their own pace and convenience.

At the end of the term, a short qualitative survey was conducted to gather feedback on how the AR application affects the perceived student's learning process. The questions in the survey used 5-point scale (1 strongly disagree, 5 strongly agree), and the results are summarized in Table 1. In general, the students had positive response toward the AR application. They found it interactive and helpful

Table 1 Summary of the students' feedback

Questions	Mean	Standard deviation
The AR application gives me better visualization on the working of distillation and to better understand different components that make up a distillation	3.50	0.55
The AR application is a useful tool to complement the lectures	3.50	0.55
The AR application increased my interest and motivation in studying and self-exploring the subject of distillation	3.50	0.55
The AR application helps me understand the relationship between distillation operating parameters and the column performance	3.83	0.75
The simulation and the assessment section are engaging enough to keep me trying to solve the problem	3.17	0.75

to better understand the topic of the class and to draw correlations between distillation parameters and the column performance. It also increased their interest and motivated them to further learn and explore the subject. In addition, they wished to implement the AR for other subjects that involve abstract topics, such as thermodynamics, fluid mechanics, and heat transfer.

With the initial development of the AR prototype, we have achieved a milestone to build up a new capability and asset in developing AR for Chemical Engineering curriculum at Nanyang Technological University Singapore. This will set a foundation for the School to further upgrade the application or to expand its development for other subjects, which is in line with the University's vision to promote technology-enhanced learning. The use of AR application allows students to self-evaluate and self-assess the distillation concepts they learned in class through the simulation and assessment scenes embedded in the AR application. Though it is not reflected quantitatively in terms of pre-test and post-test scores, students indicated that the application helped improve their learning process and increase their interest in unit operations, as shown in questionnaire results.

4 Conclusion and Future Works

A prototype of a mobile AR application for continuous distillation column has been developed and was introduced for the first time in Chemical Engineering curriculum at Nanyang Technological University Singapore. Being the initial prototype, it incorporated rudimentary principles of distillation process and was primarily used to provide visualization to enhance students' learning. There are only three parameters (reflux ratio, boil-up ratio, feed location) that can be varied as the number of trays and the feed condition were fixed. Given the simplicity of the current prototype, there is a lot of room for improvement both on the technical aspects of software and pedagogical design to enhance the seamless integration of the AR application in the course curriculum.

To improve the functionality of the application, the future works will include adding more features and variables, such as the number of trays and the feed composition. The user interface will also be improved to enhance students' engagement and participation by converting AR to VR environment. On the pedagogical aspect, assessment strategy and methods to measure the effectiveness of the application will be improved and implemented for the future cohorts of students, e.g., by giving pre-test and post-test questionnaire combined with collaborative learning. Overall, despite being in its infant stage, this initiative has shown the great potential of augmented reality as an effective tool for educational simulation and teaching.

Acknowledgements The authors would like to acknowledge the School of Chemical and Biomedical Engineering for the funding support, Siti Faatihah Binte Mohd Taib and Xu Mengshan from the School of Mechanical and Aerospace Engineering at Nanyang Technological University Singapore for working on the assessment section of the AR application.

References

1. Cai, Y., Goei, S.L., Trooster, W.: Simulation and serious games for education. Springer, Singapore (2017)
2. Cai, Y., van Joolingen, W., Walker, Z.: VR, Simulations and Serious Games for Education. Springer, Singapore (2019)
3. Martín-Gutiérrez, J., Fabiani, P., Benesova, W., et al.: Augmented reality to promote collaborative and autonomous learning in higher education. Comput. Hum. Behav. **51**, 752–761 (2015)
4. Hernandez-de-Menendez, M., Diaz, C.E., Morales-Menendez, R.: Technologies for the future of learning: state of the art. Int. J. Interact. Des. Manuf. **14**, 683–695 (2020)
5. Arvanitis, T.N., Petrou, A., Knight, J.F., et al.: Human factors and qualitative pedagogical evaluation of a mobile augmented reality system for science education used by learners with physical disabilities. Pers. Ubiquit. Comput. **13**, 243–250 (2009)
6. Klopfer, E., Squire, K.: Environmental Detectives—the development of an augmented reality platform for environmental simulations. Educ. Tech. Res. Dev. **56**, 203–228 (2008)
7. Kerawalla, L., Luckin, R., Seljeflot, S., et al.: "Making it real": exploring the potential of augmented reality for teaching primary school science. Virtual Reality **10**, 163–174 (2006)
8. Squire, K., Klopfer, E.: Augmented reality simulations on handheld computers. J. Learn. Sci. **16**, 371–413 (2007)
9. Penn, M., Ramnarain, U.: South African university students' attitudes towards chemistry learning in a virtually simulated learning environment. Chem. Educ. Res. Pract. **20**, 699–709 (2019)
10. Sahin, D., Yilmaz, R.M.: The effect of Augmented Reality Technology on middle school students' achievements and attitudes towards science education. Comput. Educ. **144**, 103710 (2020)
11. Sheik-Ali, S., Edgcombe, H., Paton, C.: Next-generation virtual and augmented reality in surgical education: a narrative review. Surg. Technol. Int. **35**, 27–35 (2019)
12. Joda, T., Gallucci, G.O., Wismeijer, D., et al.: Augmented and virtual reality in dental medicine: a systematic review. Comput. Biol. Med. **108**, 93–100 (2019)

13. Noghabaei, M., Heydarian, A., Balali, V., et al.: Trend analysis on adoption of virtual and augmented reality in the architecture, engineering, and construction industry (2020). https://doi.org/10.3390/data5010026
14. Mariscal, G., Jiménez, E., Vivas-Urias, M.D., et al.: Virtual reality simulation-based learning. Educ. Knowl. Soc. (2020). https://doi.org/10.14201/eks.20809
15. Papanastasiou, G., Drigas, A., Skianis, C., et al.: Virtual and augmented reality effects on K-12, higher and tertiary education students' twenty-first century skills. Virtual Reality **23**, 425–436 (2019)
16. Chi, M.T., Wylie, R.: The ICAP framework: Linking cognitive engagement to active learning outcomes. Educ. Psychol. **49**, 219–243 (2014)
17. Cerqueira, C., Kirner, C.: Developing educational applications with a non-programming augmented reality authoring tool. In: Amiel, T., Wilson, B., (eds.) Proceedings of EdMedia 2012-World Conference on Educational Media and Technology, pp. 2816−2825. Association for the Advancement of Computing in Education (AACE), USA, Denver, Colorado (2012)
18. Schofield, D.: Mass effect : a chemical engineering education application of virtual reality simulator technology. J. Online Learn. Teach. **8**, 63–78 (2012)
19. Wu, W., Hartless, J., Tessei, A., et al.: Design assessment in virtual and mixed reality environments: comparison of novices and experts. J. Constr. Eng. Manag. (2019). https://doi.org/10.1061/(ASCE)CO.1943-7862.0001683
20. Knierim, P., Kiss, F., Schmidt, A.: Look inside: understanding thermal flux through augmented reality. In: Adjunct Proceedings of the 2018 IEEE International Symposium on Mixed and Augmented Reality, pp. 170–171. IEEE, New York (2018)
21. Augmented reality lets students operate a chemical plant: Available from: https://www.rochester.edu/newscenter/augmented-reality-chemical-plant-297792/ (2018)
22. Williams, A.J., Pence, H.E.: Smart phones, a powerful tool in the chemistry classroom. J. Chem. Educ. **88**, 683–686 (2011)
23. Bulagang, A.F., Baharum, A.: Mobile-augmented reality framework for students self-centred learning in higher education institutions. In: Alfred R et al (Eds.) Computational Science and Technology. Springer, New York (2019)
24. Klopfer, E.: Augmented learning : Research and Design of Mobile Educational Games. MIT Press, Cambridge, Mass (2008)
25. Rosenbaum, E., Klopfer, E., Perry, J.: On location learning: authentic applied science with networked augmented realities. J. Sci. Educ. Technol. **16**, 31–45 (2007)
26. Liu, T.Y., Tan, T.H., Chu, Y.L.: Outdoor Natural Science Learning with an RFID-Supported Immersive Ubiquitous Learning Environment. J. Educ. Technol. Soc. **12**, 161–175 (2009)

The Use of Immersive Virtual Reality Technology to Deepen Learning in Singapore Schools

Choon Guan Pang, Shanti Devi, Derick Wong, Yiyu Cai, and Ryan Kyaw Thu Aung Ba

Abstract The implementation of immersive technologies such as virtual reality is contextualised as a strategy of exploration and experimentation. This brings about changes in the instructional design of virtual environments and pedagogies of lessons that achieve the outcomes of the student-centric values-driven paradigm of Singapore's education system. Riverside Secondary School's VR hub is a futuristic virtual reality classroom that provides immersive learning through sensorial experiences. Students' feedback and results from a preliminary study validates the benefit that VR is a powerful means of visual learning to help students learn molecular and cellular representations that enhance interest and joy in learning. Students develop scientific empathy when VR give them the opportunity to actualise and be a cell or molecule as this generates a first-person perspective that can give rise to more accurate reasoning to explain scientific phenomena.

Keywords Virtual reality · Immersion · Enzymes

1 Introduction

Singapore's education system has progressed from survival-driven (1959–78), efficiency-driven (1979–96), ability-driven phase (1997–2011) and now to a student-centric, values-driven education (2012–2018) where student centricity becomes the guiding moral compass for school leaders and teachers. The student-centric, values-driven paradigm has the explicit aims of valuing the aspirations, abilities and talents of every student and developing innovative and lifelong learners [1]. In the 2016 and 2017 Singapore's Ministry of Education work plan

C. G. Pang (✉) · S. Devi · D. Wong
Riverside Secondary School, 3 Woodlands Street 81, Singapore, Singapore
e-mail: pang_choon_guan@moe.edu.sg

Y. Cai · R. K. T. A. Ba
School of Mechanical and Aerospace Engineering, Nanyang Technology University, Singapore, Singapore

Y. Cai et al. (eds.), *Virtual and Augmented Reality, Simulation and Serious Games for Education*, Gaming Media and Social Effects,
https://doi.org/10.1007/978-981-16-1361-6_5

seminars, the Minister for Education articulated the need for cultural shifts in schools to promote joy of learning, nurture entrepreneurial dare in students and focus on values and character education [2, 3]. At the national level, there is a concerted effort to shift the educational discourse away from performative measures such as international PISA scores to the quality of learning that takes place in classrooms. For example, Singapore's Ministry of Education has launched the Personalised Digital Learning Programme (PDLP) whereby every secondary school student would own a learning device by end 2021. The learning device is to be used in tandem with the national e-learning platform, the Student Learning Space (SLS) and other educational technologies to personalise and enhance students' learning (see Fig. 1).

The current efforts to refine the educational system's adaptive process towards more exploration rather than exploitation may be seen as a form of rhetoric to shift the performative discourse away from plasticity to authenticity in state actors. From a strategic planning perspective, however, there is a question of whether such a change will be sustained over the course of time or will it be just another variant of the organisational culture in the short term. The basic problem is that adaptive mechanisms are myopic [4, 5]. The current forms of organisational planning, management and learning tend to privilege outcomes, threats, opportunities and preferences in the temporal and spatial neighbourhoods of an adaptive agent [6]. The dynamic flux in the social context of schools in Singapore has an impact on the mutual learning and development and diffusion of organisational knowledge in managing the trade-off between exploitation and exploration as well as balancing policy technologies of reason and technologies of foolishness.

It would be easier for the schools to experiment with variability through exploration in the short term through superficial manifestations such as playing devil's advocate and brainstorming sessions. Herbert Simon's conception of goal-less design provides a way forward for the teacher to proceed with exploring new consequences and new solutions unrelated to the original goal that may later turn out to be beneficial or intellectually stimulating [7]. A teacher who embraces

Fig. 1 Student using a personalised digital learning device

goal-less design may thus find himself enjoying the design process and discovering new knowledge and benefits for learners while preparing for lessons. This teacher, for the temporal moment, reconnects with his soul and its passions, and finds joy and satisfaction. A supportive environment that encourages teachers to pursue goal-less design goals will empower them to make a difference in the students' lives holistically with passion and purpose. March's [8] suggestion of the technology of foolishness as a counter to the policy technology of reason re-balances the optimisation agenda based on models of rationality with a "satisficing" approach that will buffer the teacher's soul from the pressures of performativity. March argues that educators should aim to cultivate in students the beauty of knowledge rather than how useful knowledge is, and as such should be guided by the logic of appropriateness rather than the logic of consequences.

If the current educational system is indeed operating from a position of strength due to evolutionary gains in efficiency and quality of central planning and resource support through the years, it may be argued that it is now time for the system to make systemic shifts and be bold in truly changing the discourse and culture in schools to pursue revolutionary change through the policy technologies of foolishness. A commitment to rationality as an instrument of exploration and foolishness, not the other way round, might enable the educational system to navigate better in a complex environment. This will set free the professional's soul as teaching is ultimately a design activity and a discipline [9] and not a science of rationality, and hence would thrive on a supportive organisational culture anchored on policy technologies of foolishness.

With digital technology being a key driver of growth, there is much scope for immersive technologies to be used as learning tools in schools and hospitals [10]. The emergent new economy required Singaporeans to develop new skillsets such as innovation, flexibility, entrepreneurship, creativity and a commitment to lifelong learning [11]. This is a praxis of March's [8] seminal ideas of cultivating a technology of foolishness in organisations to help foster an open culture in schools towards experimentation, particularly in the aspect of digital transformation. The easy availability and "plug and play" feature of most, if not all of the modern virtual reality (VR) technologies such as Oculus, HTC Vive headsets or Samsung Gear VR, Google Cardboard and a wide range of apps and software, means that immersive technologies have really come of age and it is now in the hands of schools and consumers to make the move. To paraphrase the words of British essayist and poet T.S. Elliot, we must "risk going too far (for our students) to find out how far we can go".

The launch of the Singapore Teaching Practice (STP) represents a key step that incorporates the immense possibilities of learning when teachers make the move to engage and motivate students in the current educational paradigm [12]. The STP is based on the Singapore Curriculum Philosophy which is in turn grounded on theories of social constructivism. The teaching areas are grounded on the idea of cognitive constructivism, first proposed by Piaget [13], that believes that "knowledge arises through a process of active construction" [14] and "how we all make sense of our world" [15]. The Singapore Curriculum Philosophy places the child's

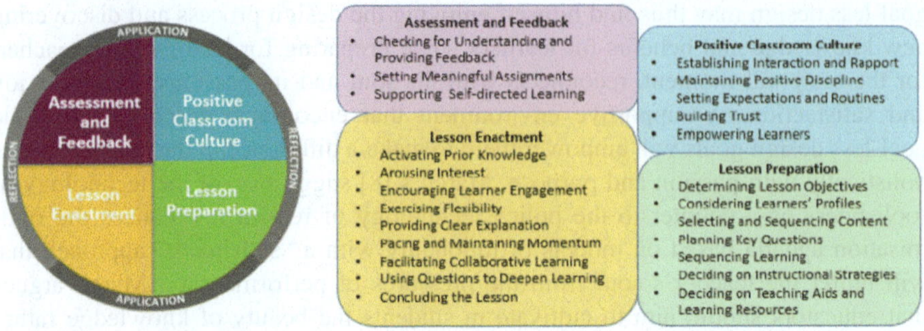

Fig. 2 Singapore Teaching Practice, 4 teaching processes and 24 teaching areas

needs at the centre when designing learning experiences and believes that learning flourishes in caring and safe environments when students are given the opportunity to construct knowledge actively, and formative assessment is used to address and bridge learning gaps [12]. The four core teaching processes—Assessment and Feedback, Positive Classroom Culture, Lesson Preparation and Lesson Enactment—make explicit 24 teaching areas or competencies that teachers can tap on in terms of strategies and approaches for classroom teaching and learning. This involves deepening and situating classroom learning in collaborative, authentic and real-world contexts. Figure 2 shows a summary of the STP.

2 Immersive Virtual Reality Technology

The beginning of the development of virtual reality as immersive technology can be traced to 1965 when Ivan Sutherland in his lecture called "The Ultimate Display" envisions the use of devices to create virtual reality by stating "if the task of the display is to serve as a looking-glass into the mathematical wonderland constructed in computer memory, it should serve as many senses as possible". Sutherland [16] published a paper detailing the very first head mounted three-dimensional display that he had developed at MIT Lincoln Laboratory which was capable of detecting the users' head positions and a novel way of displaying information to the user through cathode ray tubes, albeit with just 40° field of view.

The term "virtual reality" was subsequently coined by Yaakov Garb. In 1987 and in 1989, computer scientist Jaron Zepel Lanier proposed the term "virtual reality" or VR to describe the world of computers [17]. Virtual world, on the other hand, is described as "a computer-based, immersive, 3D multi-user environment that simulates real life, experienced through graphical representation of the user", of which *Second Life* is a notable example. To date, many definitions exist for VR, such as VR being a medium in which the user is effectively immersed in a responsive virtual world [15] and a type of simulation in which computer graphics are used to create a realistic-looking world [18]. Heim [19] states that 3 *Is*—*Immersion, Interactivity and Information Intensity*—are characteristics of VR systems.

VR is an immersive media experience that replicates the real world and allows users to interact with this world in ways that feel as if they are there. The immersive nature of VR in education occurs as the psychological processes in VR are similar to the psychological processes when people construct knowledge through interaction with objects and events in the real world [20]. Immersion is the subjective impression that one is participating in a comprehensive, realistic experience [21]. Immersion in a digital experience involves the willing suspension of disbelief, and the design of immersive learning experiences that induce this disbelief draws on sensory, actional and symbolic factors [21].

Immersion in educational settings can enhance learning in three ways—enabling multiple perspectives, situated learning and transfer of learning. Dede [21] elaborates that immersive interfaces can foster educational experiences-based situated learning or legitimate peripheral participation [22], where learners interact with virtual entities and/or peers in authentic contexts, activities and assessment. In other words, the pedagogy of using an immersive technology such as virtual reality must be situated in real-world environments or problems, and provide as much opportunities for students to work with each other collaboratively, with the teacher acting as an active facilitator and mentor in a supportive and caring environment.

In a River City project, research project funded by National Science Foundation (NSF) in the USA, such an immersive, authentic and supportive environment, has been found to produce more substantial gains in knowledge and skills in scientific inquiry and the students are able to develop problem-solving skills compared to conventional instruction. Students regardless of gender, ethnicity or ability level have also reported higher engagement, and hereby students behave like scientists as they collaboratively immerse themselves inside a simulated, historically accurate nineteenth-century city and solve a real problem of why people are getting sick and what actions can be taken to remove the illness [21]. The research indicates that active learning based on immersive situated experiences coupled with frequent opportunities for reflection is both motivating and powerful for a broad spectrum of high- and low-performing students.

3 Situated Learning—A Social-Constructivist Theory of Learning

Vygotsky [23] is one of the first scholars to articulate a social-constructivist view of learning based on the premise that knowledge is co-constructed and influenced by the surrounding environment—culture and context. According to Lave and Wenger [22], Vygotsky's concept of the zone of proximal development (ZPD) has received various interpretations which fall into three categories—firstly, ZPD can be characterised as the learning gap between novice and experienced problem solvers, which can be bridged through appropriate pedagogical interventions; secondly, ZPD can also be the distance between cultural knowledge that is made accessible through instruction and everyday experiences; thirdly, ZPD is a distance that consists of a collectivist or societal dimension that takes into account the social world in its analysis. It is from this third perspective of ZPD that Lave and Wenger [22] first proposed the idea of situated learning, or legitimate peripheral participation to describe learning as social practice, or participation in communities of practice, in contrast to the traditional view of learning as internalisation and acquisition of propositional knowledge.

Lave and Wenger's theory of situated learning states that learning is a social process whereby knowledge is co-constructed and implies that classroom context is the key to shaping the quality of learning. They advocate that learning is in the relationships between people, in short, collaboratively instead of individually, shaped by the learner participating as members of a sociocultural community, or what is being termed, as a community of practice. The nature of situated learning can best be described as understanding and experience, contemplation and involvement being in constant interaction in a learning curriculum [22]. A learning curriculum is a field of learning resources viewed from the perspective of learners. Thus, a situated learning curriculum would contain elements such as varying interests of members making diverse contributions and perspectives to problem-solving or some other cognitive activities. Participation in a community of practice can be viewed as a set of relations among persons, activity and world, or context. The 3Cs—community of practice, sociocultural context and collaborative activities—are present in a situated learning curriculum. Figure 3 shows a summary of the key elements of a situated learning curriculum.

Fig. 3 Situated learning curriculum [22]

Because situated learning theory is essentially a construct of learning as social practice, learning activities, tasks and technologies are avenues and means to build social communities or structures in learning. In addition, while pedagogy is important to a teacher's effectiveness in the classroom, pedagogy is not the source of learning, and rather learning takes place through the participation in a community of learning. In the classroom context, this community of practice consists of three actors—the teacher, the learners and the other students, as members of the community. The teacher's role is to facilitate learning by providing a variety of experiences and opportunities for learners to explore and experience and by doing so encourages the learner's new understandings [22].

4 Instructional Design of Immersive Virtual Reality Technology

Various researchers have tried to classify the types of implementation of virtual reality. Tice and Jacobson [24] have proposed three types—immersive VR, which requires the user wearing a headset, desktop VR and third-person VR. Moore [25] also suggested three types of implementation—sensory-immersive VR, desktop and text-based VR. A technology like Google Cardboard can be considered as entry-level immersive VR with some features of a desktop VR. The evolution of VR implementation to immersive experiences parallels the development of instructional design of technological resources to support learning.

According to Moore [25], there are four generations or stages of instructional design development corresponding to the development of educational theory.

- 1st stage—traditional drill and practice based on behaviourist theory developed which focuses on imparting objective knowledge or content to the learner.
- 2nd stage—focuses on the designer and strategies he or she may use to reduce the cognitive load on students and thereby facilitate instruction.
- 3rd stage—focuses on the relationship between the user and the information, with some adaption to individual learning styles built into the program that responds to user interaction.
- 4th stage—focuses on the constructivist premise that knowledge is constructed by the learner.

The development of an immersive virtual reality software by Nanyang Technological University team to teach the lock-and-key interaction of enzymes [26], which is a crucial topic for students to learn in the biology high-school curriculum, is a case in point of a fourth-generation instructional design strategy. Refer to Fig. 4 for a diagram of the software. Students are placed in a virtual laboratory environment and then use their sense of touch and sight to bring enzyme and substrate together according to the lock-and-key hypothesis.

Fig. 4 Constructivist instructional design in development of enzymes software

In the instructional design of virtual reality (VR) for educational purposes, Cai [27] proposes the term "in-depth learning" to describe the goals of technologies such as VR in education. Cai [27] explains that 3D visualisation can assist students to better understand learning objects, concepts and processes as visualisation is traditionally 2D-based, whether it be in textbooks or projection screens. He further posits that interactions between users and virtual objects in virtual worlds, such as a person being in a living virtual cell and observing its organelles, or a 3D-enabled simulation-based visualisation of river formation or an aircraft engine, would provide realistic and dynamic processes to enhance learning. Students can understand difficult concepts and develop scientific empathy. When a person puts on a immersive VR set and become a aged or homeless avatar, his consciousness is transferred to the avatar and he develops empathy for the aged or homeless as a result of the experience. Likewise, when a student uses VR to learn about cells and molecules, he transfers his consciousness to the molecule or cell being studied and as a result, he is able to intentionally actualise his learning in a comprehensive manner. To quote a student, after she had immersed herself in the VR system, she became a cell and she said that she felt so small. This generates a first-person perspective as the subject becomes a cell or molecule. Scientific empathy is developed when a person is able to recognise and understand the first-person perspective on a situation. Using the example of VR enzyme software, the person enters into the threshold of the world of enzymes and molecules when he holds and manipulates the molecule. VR is so immersive that one forgets the chasm between the physical and virtual and his physical existence merges with the virtual. For example, in a secondary 2 lesson on nutrients and digestion, students participate in hands-on activities to learn about nutrients and how enzymes work in digestion through VR. The teacher first demonstrate the enzyme VR software to explain the lock-and-key hypothesis. Students then take turns to experience the enzyme VR software or watch 360 videos of the digestive system. Figure 6 illustrates the lesson design map used by teachers to plan for VR lessons that is based on the Student Learning Space (SLS) online platform.

In-Depth Learning Framework (to deepen learning through Immersive & Interactive Learning)			
3D enabled Visual Learning	3D enabled Simulation-based learning	3D enabled Constructivist Learning	3D enabled Engaged Learning

Fig. 5 In-depth learning framework

5 Riverside Secondary VR Project

Riverside Secondary School's VR project is an interdisciplinary project that involves the integration of disciplines of ICT, national education and biology to develop lessons in the teaching of cells, viruses and enzymes that closely integrate the affordances of virtual reality and its ability to increase understanding of the real world through immersive experiences. The resulting interdisciplinary product is situated in a three-dimensional virtual world that can help students to visualise the microcosmic cellular environment, manipulate its organelles and conduct a virtual laboratory experiment to demonstrate enzyme–substrate interactions. Students gain fresh perspectives when they are able to hold and feel cell organelles like mitochondria in their hands. See Fig. 7 of students in action.

Students collaborate with each other to discover the epidemiology of the SARS and COVID-19 coronaviruses and suggest ways to prevent its spread of mutant flu viruses. Students manipulate the enzyme molecules and substrate molecules to understand more about the lock-and-key hypothesis that is so foundational in studies of human digestion.

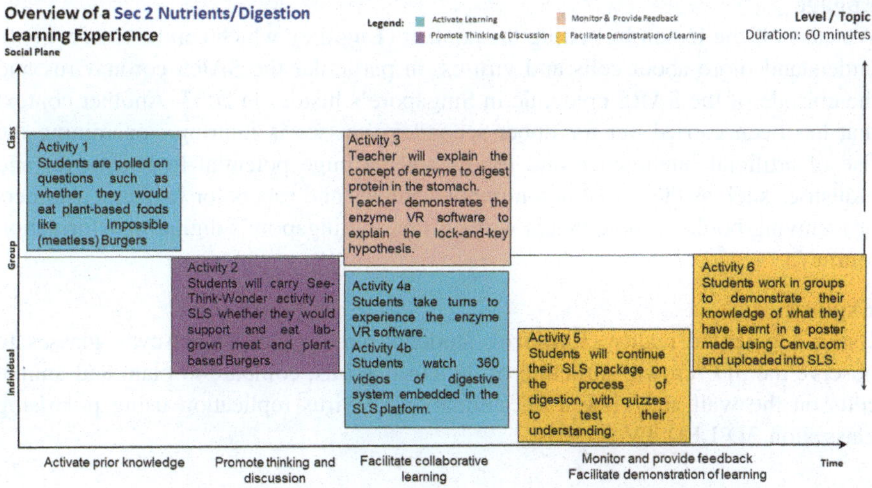

Fig. 6 SLS lesson design map for VR lesson on nutrients and digestion

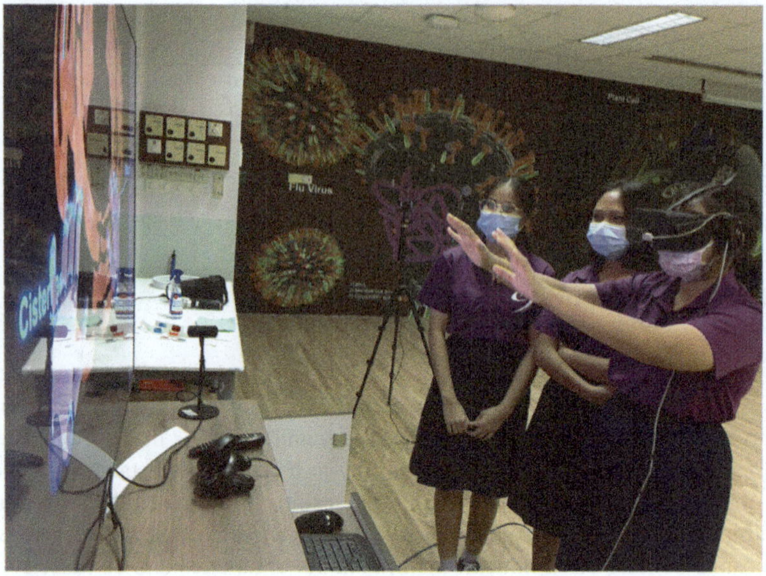

Fig. 7 Students experimenting and learning in the cells VR software

The interdisciplinary approach to learning is more consistent with constructivist philosophy, where knowledge is constructed by the learners as they immerse themselves in the VR/AR learning environment. This is situated learning at work as the learning context plays a great part in the learners' cognition and understanding. A lesson plan was developed based on the 5E guided inquiry model—Engage, Explore, Explain, Expand and Evaluate (Fig. 8).

Engage

We start off the lesson by setting the context (**Engage**), which can be to use VR to understand more about cells and viruses, in particular the SARS coronavirus and the episode of the SARS epidemic in Singapore's history in 2003. Another context that has been carried out for upper secondary classes is futuring—examining the rise of artificial intelligence and robots and its huge potential in businesses and industries such as the use of autonomous vehicles and robots for serving customers or archiving books, a topic that is very current in Singapore's digital transformation journey.

Explore

Using augmented reality (**Explore**), students put on the 3D red-cyan glasses to observe the 3D "crown" structure of the coronavirus, compare to plant and animal cells on the wall and view a 3D animation of virus replication using polarising glasses on 3D LED TV screens.

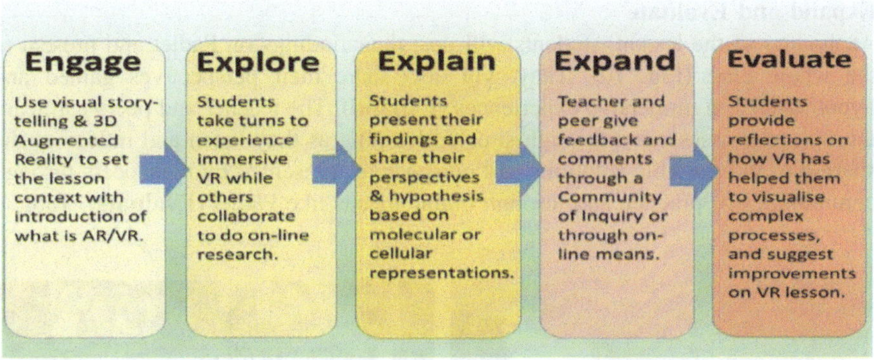

Fig. 8 Riverside's 5E inquiry learning experience model to implement VR lessons

Explain

Students will work in teams to conduct online research (**Explain**) on the problem statement or real-world situation. Examples include suggesting ways to prevent its spread of flu or respiratory viruses such as SARS coronavirus or flu viruses in a population and the benefits or pitfalls of the pervasive use of AI and robots in society.

Explore and Explain (using VR technologies concurrently)

They will take turns from the research work to experience the virtual reality software, be it the Virtual Cell module developed, and InCell or InMind apps or the Enzyme Virtual Lab, also developed. These VR experiences will help them to deepen their learning as they are able to better visualise complex 3D objects and/or interact with them through the Oculus headsets (Fig. 9) or by acting as a silent observer at the periphery.

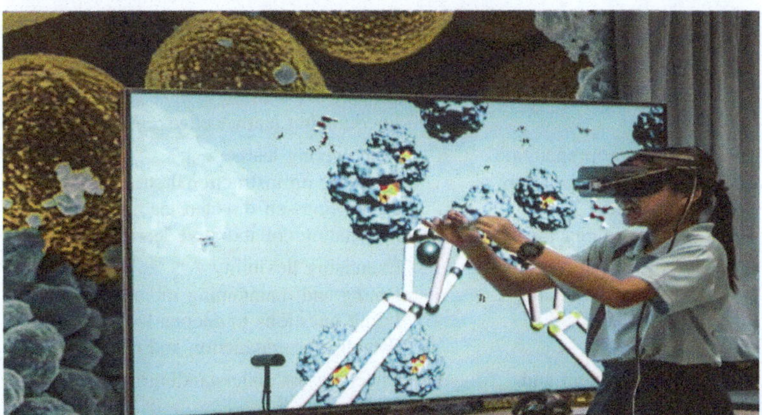

Fig. 9 Student experiencing the enzyme VR software

Expand and Evaluate

At the end of the lesson, students will post their findings on Padlet and present to the whole class (Fig. 10). They will also share their perspectives gained and hypothesis from immersive experiences (**Expand**). The teacher and peers can learn through peer assessment through situated learning as they listen and reflect on the sharing of the possible solutions and perspectives presented by their peers through communities of practice and through online means like Padlet (**Evaluate**). Table 1

Fig. 10 Students collaborating and sharing in communities of practice during a VR lesson

Table 1 Alignment of 5E model to STP processes and teaching areas

5E approach	STP teaching process(es)	STP teaching area(s) *Areas may vary depending on lesson plan*
Engage	Lesson preparation Positive classroom culture Lesson enactment	Determining lesson objectives Considering learners' profiles Activating prior knowledge Setting expectations and routines Arousing interest
Explore	Lesson preparation Positive classroom culture Lesson enactment	Empowering learners Planning key questions Selecting and sequencing content Sequencing learning Encouraging learner engagement
Explain	Lesson preparation Positive classroom culture Lesson enactment	Empowering learners Deciding on instructional strategies Supporting self-directed learning Facilitating collaborative learning
Expand	Lesson enactment Assessment and feedback	Exercising flexibility Pacing and maintaining momentum Using questions to deepen learning Establishing interaction and rapport
Evaluate	Assessment and feedback Positive classroom culture	Checking for understanding and providing feedback Building trust Concluding the lesson

shows the alignment of 5E inquiry learning experience model to the processes and teaching areas of the Singapore Teaching Practice.

6 Results of Preliminary Research

Out of a sample size of 21 secondary three students, the majority are positive and receptive about the use of VR in enhancing their learning in curriculum lessons. Eighty percentage of the students answered SA/A (Strongly Agree/Agree), with 50% SA on the question of "I hope that the school would give students more opportunities to learn and experiment with digital technologies such as VR to explore the real world in 3D and understand its dynamic interactions and concepts." In a related question, 80% of students answered SA/A, with 47% SA on the question of "VR is an exciting and futuristic ICT tool that can improve learning in science and other subjects".

On the general use of ICT for learning during curriculum by teachers, 61% of students answered SA/A, with only 19% SA on the question "I enjoy curriculum lessons more if the lesson is ICT-based involving peer collaboration using iPads/laptops." Seventy-one percentage responded with SA/A, with only 9% SA on the question "I am more motivated to learn when my teacher makes an effort to use more ICT tools and engage us through collaborative activities".

A short survey is conducted to obtain empirical research data to find out the efficacy of learning with VR on the topic of enzymes on the students' test scores performance. The quiz on enzymes consisting of 2 multiple-choice and 3 structured questions and totalling 10 marks was administered to students from two classes of 3/5 and 3/6 about 2 weeks after the class 3/5 went through the VR lesson.

The hypothesis is that going through the VR lesson on enzymes has helped the students in 3/5 to visualise and understand the complex interactions between enzymes and substrates under the lock-and-key hypothesis and the effect of environmental factors such as pH on the interactions and the enzyme activity levels.

7 Discussion and Future Research

The average test scores are 8.2 upon 10 for class 3/5 students who had gone through the VR lesson on enzymes (experimental group) and 6.875 upon 10 for class 3/6 who did not go through the VR lesson (control group). Using the equation below, we can obtain the effect size of the experimental group. The standard deviation refers to that of the control group.

$$\text{Effect size} = \frac{\text{Mean of experimental group} - \text{Mean of control group}}{\text{Standard deviation}}$$

This works out to an effect size of 0.50 for the experimental group. An effect size is exactly equivalent to a "Z-score" of a standard normal distribution. For example, an effect size of 0.8 means that the score of the average person in the experimental group is 0.8 standard deviations above the average person in the control group and hence exceeds the scores of 79% of the control group [28]. Hence, this effect size of 0.50 indicates the VR has a positive effect on learning of the molecular representations of enzymes and their interactions.

This preliminary study suggests the possible impact of immersive technologies such as VR to enhance interest and motivation, as well as achievement in the learning of science for certain topics such as enzymes in the high-school curriculum. Further research should be done to probe how the use of immersive technologies such as VR can be used in tandem with personalised learning devices at the secondary and high school levels.

Dalgarno and Lee [29] summarised the ten characteristics, affordances and benefits of three-dimensional virtual environments (3D VLE) and used it as a framework to suggest further research directions in the areas of interactive and immersive virtual worlds. Their framework serves as a useful guide for future research in the field of immersive technologies such as virtual reality.

Dalgarno and Lee [29] pointed out that sound instructional design principles and research-informed models and frameworks of instructional design should be considered when designing 3D VLE and, by extension, immersive technologies such as virtual reality for education. They aptly highlighted the need for developers/researchers and teacher practitioners to come together to dialogue and discuss the why, what and how of pedagogic strategies and tools that will best support teachers in implementing such technologies in the classroom. For instance, a team of teachers in Riverside Secondary School has been working closely with NTU's team in the development and piloting of virtual and augmented reality software in learning since 2017.

References

1. Heng, S.K.: Keynote Address, Minister for Education, at the Ministry of Education Work Plan Seminar, on Wednesday, September 12 at Ngee Ann Polytechnic Convention Centre. https://www.moe.gov.sg/news/speeches/keynote-address-by-mr-heng-swee-keat–minister-for-education–at-the-ministry-of-education-work-plan-seminar–on-wednesday–12-september-2012-at-920-am-at-ngee-ann-polytechnic-convention-centre (2012). Accessed 4 April 2018
2. Ng, C.M: Welcome Address by Minister of Education (Schools) at the 2018 International Congress for School Effectiveness and Improvement (ICSEI) (2018)
3. Ng, C.M.: MOE FY 2017 Committee of Supply Debate Speech by Minister of Education (Schools), at Singapore Parliament (2017)

4. Denrell, J., March, J.G.: Adaptation as information restriction: the hot stove effect. Organ. Sci. **12**, 523–538 (2001)
5. Levinthal, D.A., March, J.G.: The Myopia of Learning. Strateg. Manag. J. **14**, 95–112 (1993)
6. March, J.G.: Rationality, foolishness, and adaptive intelligence. Strateg. Manag. J. **27**, 201–214 (2006)
7. Chua, S.M.J.: In praise of folly: on seriously playful curriculum design. Educ. Today **58**(4), 18–23 (2008)
8. March, J.G.: A Premier on Decision Making: How Decisions Happen. The Free Press (1994)
9. Cross, N.: Designerly ways of knowing: design discipline versus design science. Design Issues **17**(3), 49–55 (2001).https://doi.org/10.1162/074793601750357196
10. Tan, W.Z.: Schools, hospitals can pave way for use of VR as learning tools. TODAY online. https://www.todayonline.com/singapore/use-vr-education-clinical-training-being-explored (2017). Accessed on 11 Nov 2018
11. Saravanan, G.: Globalisation, the singapore developmental state and education policy: a thesis revisited. Glob. Soc. Educ. **5**, 53–70 (2007). https://doi.org/10.1080/14767720601133405
12. MOE.: The Singapore Teaching Practice. https://www.moe.gov.sg/about/singapore-teaching-practice (2017). Accessed 30 Oct 2018
13. Piaget, J.: Piaget's theory. In: Mussen, P. (ed.) Handbook of Child Psychology. vol. 1, 4th edn. Wiley (1983)
14. Mascolo, M.F., Fischer, K.W.: Constructivist theories. In: Hopkins, B., Barr, R.G., Michel, G., Rochat, P. (ed.) Cambridge Encyclopedia of Child Development. pp. 49–63. Cambridge University Press (2005)
15. Brooks, M.G., Brooks, J.G.: The courage to be constructivist. Educ. Leadership **57**(3). https://www.ascd.org/publications/educational-leadership/nov99/vol57/num03/The-Courage-to-Be-Constructivist.aspx (1999). Accessed 10 Nov 2018
16. Sutherland, I.E.: AFIPS Proceedings of the Dec 9–11 Fall Joint Computer Conference Part I, 757–764 (1968)
17. Sala, N.M.: Virtual Reality and Education: Overview Across Different Disciplines (2016)
18. Burdea, G.C., Coiffet, P.: Virtual reality technology. Presence **12**, 663−664 (2003). https://doi.org/10.1162/105474603322955950
19. Heim, M.: The Metaphpysics of Virtual Reality, pp. 109–128. Oxford University Press (1993)
20. Winn, W.D.: A conceptual basis for educational applications of virtual reality. (HITL Report No. R-93-9). University of Washington, Human Interface Technology Laboratory, Seattle, W.A. (1993)
21. Dede, C.: Immersive interfaces for engagement and learning. Science **323**(66), 66–69 (2009)
22. Lave, J., Wenger, E.: Situated Learning: Legitimate Peripheral Participation. Cambridge University Press (1991)
23. Vygotsky, L.S.: Mind in Society: The Development of Higher Psychological Processes. University Press (1978)
24. Tice, S., Jacobson, L.: VR in visualization, animation, and entertainment. In: Jacobson, L. (ed.) CyberArts: Exploring art and Technology. Miller Freeman (1992)
25. Moore, P.: Learning and teaching in virtual worlds: implications of virtual reality for education. Aust. J. Educ. Technol. **11**(2), 91–102 (1995)
26. Ba, R., Xie, Y., Zhang, Y., Taib, S.F.B.M., Cai, Y., Walker, Z. et al.: Virtual reality enzymes: an interdisciplinary and international project towards an inquiry based pedagogy. In: Cai, Y. et al. (ed.) VR, Simulation and Serious Games for Education, pp. 45–54. Springer, Singapore (2018)
27. Cai, Y. (ed.): 3D Immersive and Interactive Learning. Springer, (2012)
28. Coe, R.: It's the effect size, stupid. Paper Presented at the Annual Conference of the British Educational Research Association, University of Exeter, England, 12–14 Sep 2002. https://www.leeds.ac.uk/educol/documents/00002182.htm (2002). Accessed 24 Aug 2018
29. Dalgarno, B., Lee, M.: What are the learning affordances of 3-D virtual environments? Br. J. Edu. Technol. **41**(1), 10–32 (2010)

Learning Life Skills Through Gaming for Children with Autism Spectrum Disorder

Guorong Hoe, Qi Cao, Jieqiong Chen, and Yiyu Cai

Abstract Children with Autism Spectrum Disorder (ASD) have social and communication impairments. Various efforts are made to assist children with ASD in improving their quality of life, education, and independence. Obviously, learning is utmost important for this aim. This paper describes game-based life skills learning for children with ASD. A virtual reality (VR) based serious game has been designed to aid children with ASD in their learning of life skills, particularly tooth brushing and simple cooking. Control and experiment studies have been conducted in this research for the purpose to compare the effectiveness of the methods developed. Initial evaluation results are promising. Hopefully, the outcomes could benefit children with ASD to generalize the skills learnt into the real-world experience after the serious games play.

Keywords Serious games · Virtual reality · Life skills · Special needs education · Autism spectrum disorder

1 Introduction

1.1 Autism Spectrum Disorder

Global statistics show that there is an increase of Autism Spectrum Disorder (ASD) with the international prevalence rate 0.6% applied across cultures [17, 18,

G. Hoe · Y. Cai (✉)
School of Mechanical and Aerospace Engineering, Nanyang Technological University, Singapore 639798, Singapore
e-mail: myycai@ntu.edu.sg

Q. Cao
School of Computing Science, University of Glasgow, Singapore campus, Singapore 567739, Singapore
e-mail: qi.cao@glasgow.ac.uk

J. Chen
VARTEL Network, Singapore, Singapore

Y. Cai et al. (eds.), *Virtual and Augmented Reality, Simulation and Serious Games for Education*, Gaming Media and Social Effects,
https://doi.org/10.1007/978-981-16-1361-6_6

23, 37, 44], making ASD the fastest growing disorder in the world [36]. As a complex and lifelong neurodevelopmental condition, there is no cure for ASD at the moment. Provision of intervention services in early childhood would bring greater likelihood for positive outcomes over alternative intervention offered later in life [10, 22, 39]. There is a growing demand worldwide for special educational resources for children with ASD [30]. The advancement of the latest computing and engineering technologies makes it possible to offer serious games for the better learning of children with ASD. For those diagnosed with ASD, core challenges caused are deficits of social skill and social interactions, including social interaction rules, social initiation, emotional expressions and emotion recognition [8, 33], which may lead to potential problems of integration into society [25]. Besides the social skills, some other forms of difficulties affecting attention, executive function [20], general cognition, learning, speaking, reading, writing, math, concentration, organization, time management, social interactions, or speech comprehension [22] may also be associated with ASD [25]. All these may need long-term supports and help from families and society [1]. The Diagnostic and Statistical Manual of Mental Disorders, Fifth Edition (DSM-5), is a classification and diagnostic tool that defines and classifies mental disorders. According to DSM-5, ASD can be categorized into a dimensional manner whereby as long an individual has deficits in social communication and interaction with a high reactivity to sensory input or unusual interest in sensory aspects of the environment, can now also be considered under ASD [3].

Different types of technologies have been used in studies with an increasing interest of multidisciplinary research for ASD [10]. In recent years, many new developments have been reported using computing or engineering-based approaches for training, education or therapy for children with ASD [7]. In the following discussions in this sub-section and Sect. 1.2, we will look into these methods more closely.

Two categories of studies have been classified by DiPietro et al. [22] in their review paper: computer-based serious games, and robot-assisted tools including humanoid social robots, non-humanoid robots or virtual avatars. By searching in literature for publications in the period of January 2015 to January 2019, 18 most relevant studies have been identified (11 using robots, and 7 using serious games). Instead of serious games, Boucenna et al. [10] have argued that information & communication technology (ICT) and robotics should be considered as two categories of approaches. ICT approaches are fairly wide which could be further divided into four sub-categories: computer games-based interactive environments, virtual reality (VR) environments, avatars for serious games, and tele-rehabilitation for delivering rehabilitation services. However, Grossard et al. [25] have suggested three categories of approaches used supporting mechanisms for social skills training in children with ASD: (1) serious games with digital games and equipment, (2) robots for social skills training, and (3) iPad or mobile applications (apps) aiming to specific aspects of social life. Very recently, it has seen an emerging trend in the studies with the use of serious games [2, 9, 16, 28, 29, 34, 40, 41], and/or robotics approaches [4, 19, 26, 35, 45, 46].

1.2 *Serious Game and Simulation for Special Needs Education*

Serious games have been developed and used in various applications such as health care, rehabilitation, education and training [5, 22, 43]. Compared to the robotics approach, computer-based serious games are more accessible and cost-effective often requiring minimal physical hardware. For purpose of education or training in relation to children with ASD, serious games usually have both pedagogical theories and motivational principles of game design integrated into the same tool with the goal to help them acquiring new skills [42].

Immersive and interactive learning is increasingly recognized as an ubiquitous form in learning today [11, 12, 14]. There are good efforts made to develop digital aids for children with ASD to learn various skills from communication and psychomotor training, to social behaviour augmentation [21, 31]. Serious games in principle are simulations of real-world events. Education including special needs education and training have been two active areas for applications of serious games, simulations and VR [6, 13, 15, 20, 27, 32]. Caria et al. [16] describe a web-based application to help people with ASD by gaining skills to understand the concept of money as daily activities. Simões et al. [41] design a serious VR game using an Oculus Rift headset to teach children with ASD on taking public transportation. Bossavit and Parsons [9] present their serious game aiming to help young people with ASD learn Geography-specific knowledge. To assist children with ASD recognizing facial emotional expressions for joy, sadness, anger, and surprise, Almeida et al. [2] develop a role-playing serious game using Unity 3D. A virtual avatar-based serious game is introduced by Grossard et al. [24] to help children with ASD to learn adapted facial expression in specific social context. Serious games are used as an intervention to teach vocabulary words [29]. A platform has been developed to help 5–10 years old children with ASD learn emotional expression and recognition, through play in the virtual world [34]. Computer-based serious games can reduce costs to improve training and education of children with ASD with regard to overall quality of life [7].

1.3 *Objectives of this Research*

This research project aims to assist children with ASD to learn basic life skills with VR-enabled serious games. The virtual home designed emphasizes the overall interactivity and effectiveness of the game enhanced by visual and audio cues. It aims to help children with ASD generalizing the skills learnt from gaming into their real-life world experience. An experimental testing is conducted in order to assess the effectiveness of the VR-based serious game.

1.4 Organization of the Chapter

The remaining parts of the chapter are organized as follows. Section 2 presents the design of the virtual home game. Section 3 introduces the experiment test conducted in a special needs school in Singapore. The experiment results are evaluated and discussed in Sect. 4. This is followed by a conclusion in Sect. 5.

2 Design of Virtual Home Game

Children with ASD are often dependent on their parents or caretakers for their daily activities. These tasks common to normal people could be challenging to children with ASD when they are left alone. In this research project, we have designed a VR-based serious game for them to learn social life skills: particularly, tooth brushing, and making a cup of Milo drink. These two activities are part of common daily routine for most of Singaporean children. The VR-based serious game designed is used to train children with ASD with a long-term objective for independent living.

2.1 Virtual Home Serious Game

In this research project, a virtual home is designed according to a typical Singaporean apartment attached with kitchen, washroom, living room, etc. The virtual home serious game has been created for children with ASD to learn two types of activities; one for personal hygiene keeping, the other for simple cooking. In a long-term plan of this research, we will make the virtual home a realistic learning environment for children with ASD to learn all skills relevant to live at home.

2.1.1 Personal Hygiene Keeping—Brushing of Teeth

To perform the first type of activities, several virtual objects in the virtual home will be utilized. The basin in the washroom will serve as an essential area to learn tooth brushing by the children. In the VR-based serious game, the game-players can select the virtual items, such as toothbrush, toothpaste, water tap and cup with his/her virtual hand as shown in Fig. 1a. A Leap Motion sensor connected to the computer system is used to detect the actual hand motions and selection actions of the game-player. The hand motions include picking up the virtual toothbrush, squeezing the virtual toothpaste on the toothbrush, turning on the tap and holding the cup to collect water from the tap for rinsing of mouth. In the virtual home serious game, the game-player will be trained to learn and practice the entire activities step by step. Upon successful completion of the VR gameplay of the tooth brushing, a pre-recorded video will be displayed to the game-player to demonstrate the common step by steps for tooth brushing activity in the real environment setting of a typical physical washroom as shown in Fig. 1b. Through this method, the game-players can perform the comparison between the motions of the pre-recorded video in the real environment and those what he/she does in the virtual serious game.

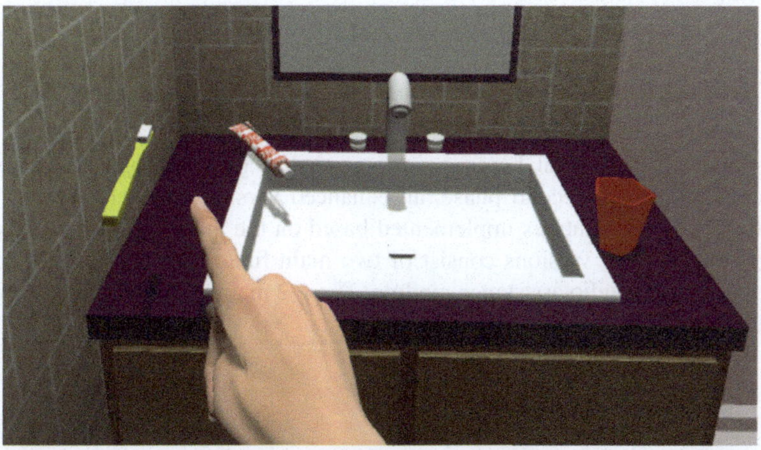

a) Interactions in the VR serious game

b) Pre-recorded video demonstration

Fig. 1 Tooth brushing activities in the virtual home serious game

2.1.2 Simple Cooking—Making of Hot Milo Drink

Milo is a popular hot drink in Singapore. As shown in Fig. 2, making a cup of hot Milo drink is implemented as a part of the designed virtual home serious game. Playing this game, children with ASD will have to consider the safety issue due to the use of hot water. To make a cup of Milo drink in the virtual home serious game, players are required to tap on the virtual items in a sequential manner accordingly: picking the virtual cup to be used for drinking, reaching out for the Milo tin to scoop the Milo powder into the virtual cup, proceeding to the virtual hot water dispenser to release the right volume of hot water into the virtual cup, and picking up a teaspoon to stir the mixture.

2.2 *Interactive Design*

The project of the virtual home serious game has gone through two phases of developments. In the second phase, an enhanced version is developed with some bugs fixed and new features implemented based on the initial version developed in the first phase. Both versions consist of two main function modules: visual assistance module and audio assistance module. These 2 main modules will be described in the sub-sections as follows.

2.2.1 Visual Assistance Module

The high quality and realistic visual effects in a VR serious game would be very helpful to deliver the training and learning contents for children with ASD [38]. In this project, the virtual home serious game has been equipped with the visual

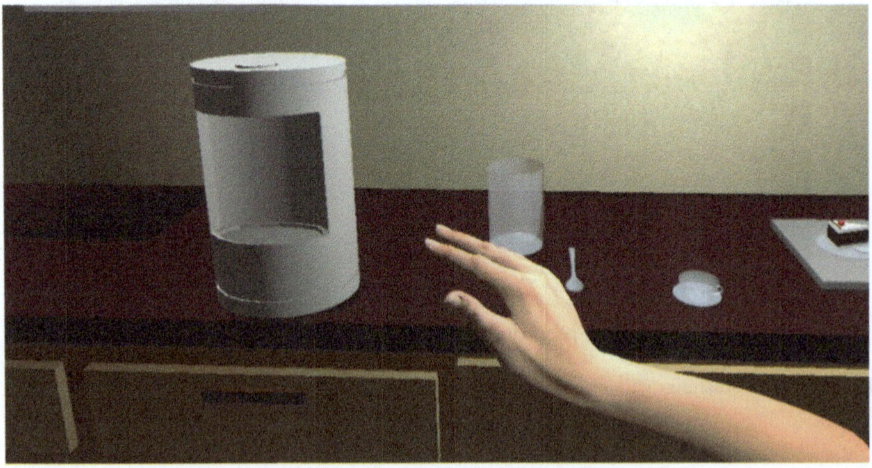

Fig. 2 Making a hot Milo drink in the virtual home serious game

assistance as supporting function. The function has been designed with these features including graphics design, visual instructions, visual feedback, animation design, and visual demonstration.

Graphics design: In the game, the virtual kitchen and the washroom are designed to closely mimic the real-world environment found in a typical Singapore apartment. The sink table in washroom and the food making table in the kitchen are designed in a great detail graphically. The colours and materials for the virtual items including decoration items such as towel, socket and showerhead are also graphically designed to replicate the real-world physical items.

Visual instructions: The contents of Graphical User Interface (GUI) and checklist are incorporated in the design to facilitate the learning process for this sequencing virtual home-based game. In the virtual home serious game, visual instructions including text words and symbols are used concurrently. The visual instructions and words can guide players to pick the right virtual items. A checklist will also appear to indicate if the action is wrong, correct, or completed successfully.

Visual feedback: To help children with ASD interact with the learning process, the feature of visual feedbacks is served instantly after a selection is made. Examples of the visual feedbacks for correct and wrong answers are shown in Fig. 3.

Animation design: This is a feature with the enhanced version only. The animation design feature is added partially according to the popular feedback received from the teachers after an initial phase development. To help children with ASD better focus on the learning process through gaming, the floating animation effects are implemented to all selected virtual items in the serious game. It is also recommended to have this animation feature assisting those children mentally handicapped or visually challenged. These children might have difficulties following the word-based instructions displayed on the screen. In the virtual home serious game, the scene will first start off statically. After the visual or audio instructions are read out, the corresponding correct item for that step will have the floating animation

Fig. 3 Example visual feedback for correct answer and wrong answers

engaged. While the rest virtual items remain at their static and original positions. Following a correct action, the next set of instructions will be read out. The next corresponding item that should be tapped on will start the floating animation. For the virtual tap in the game, water droplet effect is also added.

Visual demonstration: This is another feature implemented in the enhanced version. Upon the completion of the serious gameplays, the pre-recorded videos will be triggered to display the actual step-by-step demonstrations for conducting real-world activities of tooth brushing and Milo drink making. As introduced previously, an example of pre-recorded video for actual tooth brushing in the real environment is shown in Fig. 1b.

2.2.2 Audio Assistance Module

The audio assistance is the second main function module in the designed game. The audio assistance has embedded these features: audial instruction, background music, and audio feedback.

Audial instruction: Another feature added in the enhanced version. It has been implemented to guide the children in the learning process. Instead of simply duplicating the contents of the visual instructions, audio instructions are designed to help children to perform individual thinking in their serious gaming.

For example, the audio instruction "*Great job! Now what do you put on the toothbrush?*" will allow children to execute their own thinking process, and hence facilitate their decision process. Audio instructions are also provided if children make a wrong move, such that they will be asked to read the instructions again. Careful designs are made to ensure that different audio instructions do not overlap with one another to avoid overloading sensory feedback in children. If one action is conducted right after another, the audio instructions will be played in sequence without overlap.

Background music: In the enhanced version of the virtual home serious game, soothing and calming background music is implemented to help alleviate any possible sensory overload of the children with ASD in the form of either visually or auditory.

Audio feedback: Short audio clips are incorporated in the serious game to give instance audio feedback when an incorrect item is selected. The audio feedback feature provides the aid in directing the players to reach the correct virtual objects. This way, children can act according to the scenes in the serious game more naturally. Furthermore, upon successful completion of a learning element, i.e. tooth brushing or Milo drink making, relevant audio clips will be played to emulate the corresponding sound effects, such as brushing sounds, stirring sounds, or other sounds, respectively. The enhanced version of virtual home serious game has the refined feature with the audio feedback to better serve the learning objectives.

3 Experiment Study

The motivation and rationale behind the proposed virtual home serious games seek the benefits for the children with ASD to learn life skills by game-playing. With the designed function modules and features in the game, the interest and enthusiasm of the children could be easily reflected while the serious playing is in progress. Besides, it is crucial to measure the effectiveness of the primary learning objectives and outcomes. The purpose of evaluating the virtual home serious game is to ensure that the children achieve their learning outcomes, as what the serious game is intended to achieve.

Measurements and evaluations are done through the collection of various types of data, such as the actual game-playing data and additional test data to compare the difference between the before and after of playing the serious game. Game playing data refers to the players' in-game movement, actions and responses which are acquired by numerical variables. There are two ways to obtain gamers' playing records: one with ex situ data; the other with in situ data.

- The ex situ data refers to data that is presented from the "outside" of the game through observation. The profiles of the players such as the age, survey feedbacks, physical pre-tests and post-tests results are forms of ex situ data.
- On the other hand, the in situ data is found from the "inside" of the designed serious game. In other words, it refers to the data that can be obtained from the internal software program of the serious game [32]. With the applicable information gathered, valuable conclusions can be drawn from the assessment of the game.

3.1 The Experiment Design

3.1.1 Control and Experiment

In the current experiment of this research, a total number of 12 children from a local special needs school in Singapore participates in the study. The 12 children are divided into 3 groups, each having 4 members. *Group A* is a control group using the conventional learning method without the serious game. *Experiment Group B* is an experimental group that plays only the initial version of the proposed virtual home serious game. *Experiment Group C* is also an experimental group but they only play the enhanced version of the proposed serious game.

Experiment Group B: First experimental group playing the initial version of the developed serious game. The results and observations are recorded. *Group B* is implemented to determine whether the enhanced features in the enhanced version, when compared to the initial version of the serious game, are necessary, useful or effective in teaching the children regarding sequencing idea. It is also to determine the preferences of the children. *Group B* also gets a chance to interact with the enhanced version of the serious game after having completed the initial version. However, their results achieved with the enhanced version will not be tabulated. It can be used as references and comparison between the initial version and the enhanced version.

Experiment Group C: Second experimental group playing the enhanced version of the virtual home serious game with observations and results being recorded. The evaluation of this experimental group is the same as that of *Group B*.

For *Group B* and *Group C*, children do a warm-up to familiarize themselves with the virtual home serious game. Afterwards, the children begin progressing onto the 2 serious game scenes to perform the learning activities for the skills of tooth brushing and hot Milo drink making. They carry out each game twice, so as to ensure certain variables can be kept constant. During the process, the steps and actions performed by the children are recorded to determine if the game design and mechanisms are suitable for the children. Other information such as software bugs, errors, and accidental touching is also recorded for the comparison of 2 experimental groups.

Control Group A: This group being taught using the conventional method. Real-world objects are used for their learning. These physical objects correspond to the virtual items developed in the serious game. A real toothbrush, toothpaste, water tap and cup are used for the tooth brushing activity are shown in Fig. 4. Figure 5 shows a real Milo tin, cup, hot water dispenser, and spoon. These real items have the corresponding virtual version in the second activity, i.e. hot Milo drink making in the developed serious game.

One of the intentions of having the control *Group A* in the research project is to determine whether real-world implementation of objects would have different effects on sequential memory training, in comparison with the VR-based serious gaming contents experienced by the two experimental groups. In this sense, the conventional teaching contents of this control *Group A* cover both the classroom training environment and the home activities environment. In classroom-based teaching, real-world implementations of objects are used from time to time in schools to help children understand concepts better. In the home activities environment, family members usually guide and teach their children how to handle and interact with real objects around a household environment.

3.1.2 Experiment Sequences

Tooth brushing and Milo drink making are generally sequence-based activities. Children with ASD often have weak executive function in planning and carrying out those activities. It is therefore important to help them learn and improve their executive function. Both conventional training (control) and serious game-based learning (experiment) are able to give the players necessary items in different orders. The virtual home serious game is specifically designed for children with ASD to learn the proper order of tooth brushing and drink making, taking advantage of realistic simulation in the virtual environments. Through gaming, children

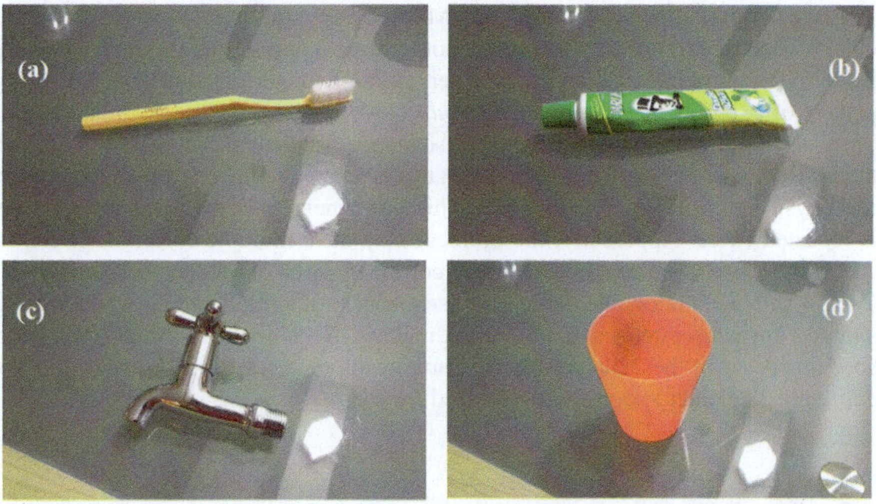

Fig. 4 Real objects used for the tooth brushing: **(a)** A real toothbrush, **(b)** a real toothpaste, **(c)** a real water tap, and **(d)** a real water cup

Fig. 5 Real objects used for Milo drink making: **(a)** A real cup, **(b)** a real Milo tin, **(c)** a real hot water dispenser, and **(d)** a real spoon

can develop their understanding on why tooth brushing and Milo drink making should have a correct sequence. For example, in the Milo drink making scene, the virtual items can be given to the gamers in any random order. They are required to identify a right sequence for their drink making with the items given.

For example, cup, water dispenser, Milo powder tin, and spoon are the basic objects used in Milo drink making for both physical and virtual training. They are given to the participants in an initial order, randomly determined. The participating players of experiment *Group B* and *Group C* will interact with the virtual objects. They are required to identify the right sequence for their drink making with the random order of objects given. Similarly, this is applied to the control *Group A* with real physical objects in a random order. This experiment process with all three groups is repeated twice. Apparently, the problem may have multiple correct answers. In such case, any answer that is considered correct in terms of sequencing is accepted. Table 1 shows 8 reasonable and correct sequences of tooth brushing while Table 2 shows 2 correct sequences of Milo drink making. Do note that normally, there is a specific spoon in the Milo tin for the participants to scoop some powder into the cup with. The other spoon should be used for stirring.

3.2 The Evaluation

Group A: In the evaluation stage, participants in this group are asked to choose real physical objects used in the conventional teaching method presented to them. The number of attempts they take before successfully identifying the correct sequence to complete both tooth brushing and Milo drink making is recorded. Each of the participants will have two chances to make their choices. The best number of attempts will be kept.

Group B: This group of participants goes through the evaluation by playing the serious game of the initial version. They are asked to choose the virtual objects in the right sequence within the virtual home environment. The number of attempts before they can successfully identify the right way of completing tooth brushing and Milo drink making will be recorded. Each of participants has two chances in their choice.

Group C: Similar to *Group B*, this group of participants go through the evaluation by playing the serious game but with the enhanced version. They are asked to choose the virtual objects in the right sequence within the virtual home environment. The number of attempts before they can successfully identify the right way to

Table 1 Reasonable sequences for tooth brushing

Cup	Tap	Toothpaste	Toothbrush
1	2	3	4
1	2	4	3
2	1	3	4
2	1	4	3
3	4	1	2
3	4	2	1
4	3	1	2
4	3	2	1

Table 2 Reasonable sequences for Milo drink making

Cup	Tap	Toothpaste	Toothbrush
1	2	3	4
1	2	4	3
2	1	3	4
2	1	4	3
3	4	1	2
3	4	2	1
4	3	1	2
4	3	2	1

Fig. 6 Objects for tooth brushing laid out in an example order (*d*) cup → (*c*) tap → (*b*) toothpaste → (*a*) toothbrush

complete tooth brushing and Milo drink making will be recorded. Each of participants also has two chances in their choice.

Figure 6 shows an example in the tooth brushing scene, the physical objects are given to the participants left to right in a sequential order of (*d*) water cup → (*c*) water tap → (*b*) toothpaste → (*a*) toothbrush. They will try to choose the physical objects in a suitable order feasible to complete the tooth brushing, e.g. (*a*) toothbrush → (*b*) toothpaste → (*c*) water tap → (*d*) water cup. Similarly, the real physical objects will be chosen in a right sequence to complete Milo drink making (Fig. 7).

4 Experiment Execution, Results and Discussion

The 12 children participating in the research are brought into the room one at a time to complete the game experiment. The experiment is conducted one-on-one. Hence there are no other distractions to the children in the experiment. The children are

Fig. 7 Objects for Milo drink making laid out in an example order (*b*) Milo powder → (*a*) cup → (*c*) hot water dispenser → (*d*) spoon

Table 3 Profiles of the 12 participants from the special need school

Child Code	Gender	Age	Profile
A-1	Male	10	Autism
A-2	Male	11	Autism
A-3	Male	8	Autism
A-4	Male	8	Autism
B-1	Male	13	Special education
B-2	Male	14	Autism
B-3	Male	14	Autism
B-4	Male	8	Autism
C-1	Male	16	Autism
C-2	Male	16	Special education
C-3	Male	15	Autism
C-4	Male	8	Autism

tasked to play the game according to the procedures laid out. The teachers of the respective children are observing at the side. At the end of the experiment session, the teachers are asked to fill in feedback forms regarding the performance of the children.

Table 3 represents the demographics of the 12 participants of the experiment. The notations of A, B and C are the corresponding group codes for the Control *Group A*, Experiment *Group B* and Experiment *Group C*, respectively.

The teachers from the special needs school help in this experiment by observing for the entire gameplay process of their children, and getting feedback from the children. Figure 8 shows one child and his teacher during the experiment. According to the feedback and observations received from the children, they enjoy this type of learning by serious gaming. In general, children like video and moving

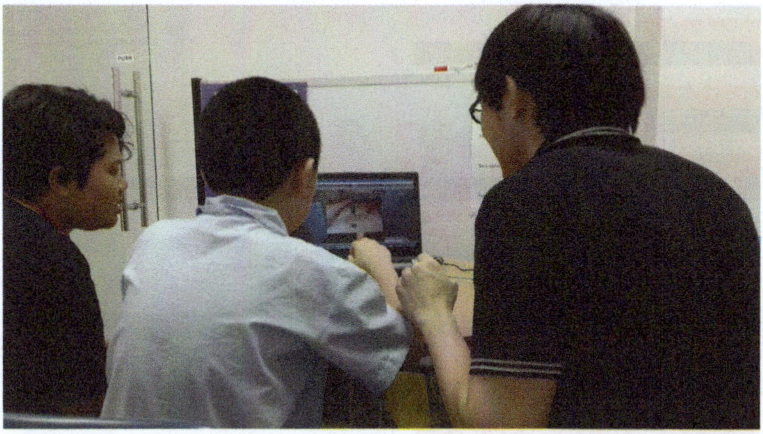

Fig. 8 A child (centre) playing the serious game with his teacher (left) observing the experiment

animations which may be helpful in generalizing of skills for them into the real world.

The experiment results for both tooth brushing and Milo drink making gameplay of these 3 groups are shown in Fig. 9. The number of attempts represents the accumulated number of actions of all the children in the Experiment *Group B*, compared to those of the Experiment *Group C*. It shows how many times a certain object or incidents, such as bugs and accidental actions, are summed up over for the 4 children in each group, before the right object for that sequence is picked up. It is observed in Fig. 9 that all children of *Group A* perform a total number of 5 times for the activity of the tooth brushing scene, while they perform a total number of 11 times for the activity of the Milo drink making scene. All children of *Group B* perform a total number of 8 times for the activity of the tooth brushing scene, and a total number of 16 times for the activity of the Milo drink making scene. While *Group C* perform the least total number of attempts, 4 times, for the activity of the tooth brushing scene, and also the least number attempts for the activity of the Milo drink making scene. Table 4 shows the average number of attempts for the control and experimental groups before making a right sequence. It can be seen from the numbers in Fig. 9 that the mistakes made by the players are significantly reduced in *Group C* compared to other groups. Children from the experiment *Group C* are able to comprehend, select and interact with the correct objects in their first tries. Thus it indicates the enhancement of the design improves the learning of children with ASD, and good instructional design helps minimize wrong or accidental selections during the VR serious gaming.

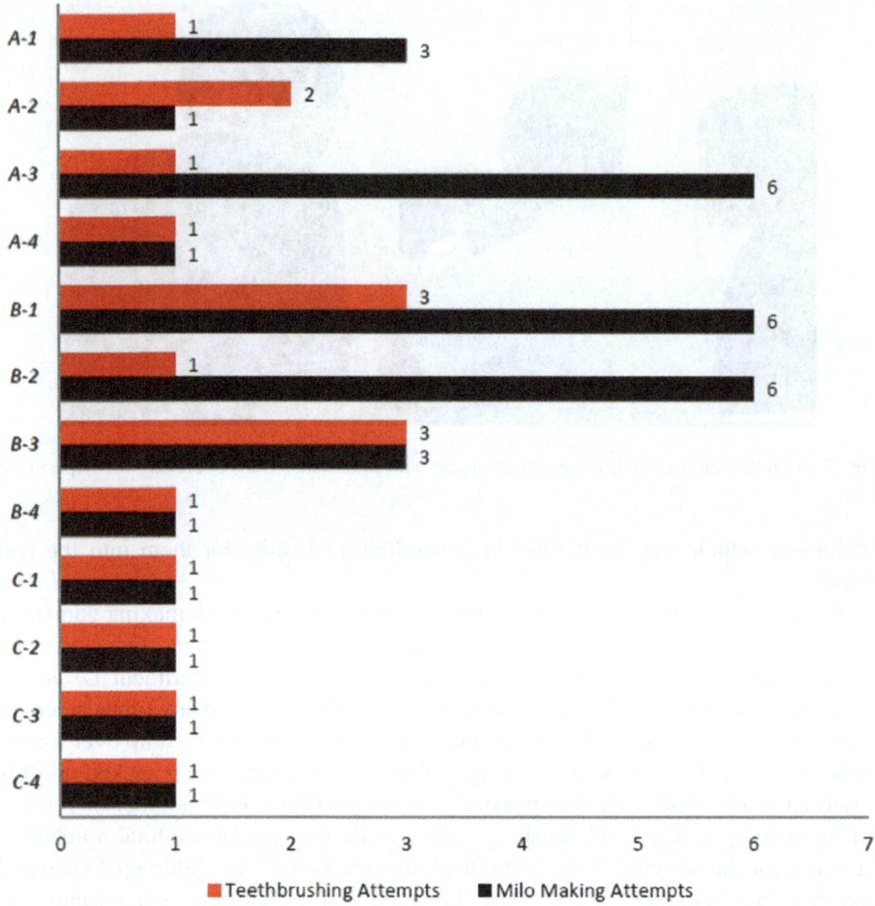

Fig. 9 Number of attempts taken for each individual child before getting the right sequence

Table 4 Average attempt numbers for control and experimental groups before making a right sequence

Group	Average attempt for tooth brushing	Average attempt for Milo drink making
A	1.25	2.75
B	2	4
C	1	1

5 Conclusions

The purpose of designing this virtual home serious games is to produce a safe and effective environment for children with ASD to learn practical life skills by using interactive gaming. In this case, parents and educators can utilize the capability of serious games for teaching children with ASD and other special needs. Through serious gaming, these children can learn from the mistakes they make. It allows consistent practice to help ensure that the children are able to execute the steps and sequences correctly in real life. Through serious gaming, the children develop their interests in life skills learning. This can result in big beneficial advantages such as players being able to have fun in their learning with the aid of input stimuli, engagement and virtual reality.

There are a lot of life skills children with ASD have to learn in their early years. This is a test-bedding project with two serious games developed. Tooth brushing and hot Milo drink making are identified after several consultations with local special needs schools. Instead of having the serious game solution replace real-life learning completely, we also believe it is an ideal approach to combine the serious gaming and real-life learning for children with ASD just like how a pilot spends a substantial amount time with flight simulator before their actual flying.

There are certain limitations with the current research. Only 12 children with ASD participate in this experiment. A larger sample size is required to achieve better results. Moreover, there is a big age gap among the 12 participants which may possibly attribute to inconsistent results. For children with ASD, serious games should have several special design considerations including GUI and natural user interfaces (NUI).

The present virtual learning environments developed are basically projection-based solutions designed for group-based or team-based learning. For individual-based learning, head-mounted displays like Oculus or HoloLens may be able to provide better functions supporting the children's learning.

Acknowledgements This project is partially supported by the Temasek Trust Funded Singapore Millennium Foundation. The authors would like to thank many people and organisations who help the project in one way or another.

References

1. Alcorn, A.M., Ainger, E., Charisi, V., Mantinioti, S., Petrović, S., Schadenberg, B.R., Tavassoli, T., Pellicano, E.: Educators' views on using humanoid robots with autistic learners in special education settings in England. Front. Robot. AI **6**, 1–15 (2019)
2. Almeida, L.M., Silva, D.P.D., Theodório, D.P., Silva, W.W., Rodrigues, S.C.M., Scardovelli, T.A., Silva, A.P.D., Bissaco, M.A.S.: ALTRIRAS: A computer game for training children with autism spectrum disorder in the recognition of basic emotions. Int. J. Comput Games Technol (2019)

3. American Psychiatric Association, Diagnostic and Statistical Manual of Mental Disorders. Fifth Edition (2013)
4. Attawibulkul, S., Asawalertsak, N., Suwawong, P., Wattanapongsakul, P., Jutharee, W., Kaewkamnerdpong, B.: Using a daily routine game on the BLISS robot for supporting personal-social development in children with autism and other special needs. In: 58th Annual Conference of the Society of Instrument and Control Engineers of Japan, Hiroshima, Japan (2019)
5. Barajas, A.O., Osman, H.A., Shirmohammadi, S.: A serious game for children with autism spectrum disorder as a tool for play therapy. In: IEEE International Conference on Serious Games and Applications for Health (SeGAH), Australia (2017)
6. Barakova, E., Gillessen, J., Feijs, L.: Social training of autistic children with interactive intelligent agents. J. Integr. Neurosci. **8**(1), 23–34 (2009)
7. Benssassi, E.M., Gomez, J.C., Boyd, L.E., Hayes, G.R., Ye, J.: Wearable assistive technologies for autism: opportunities and challenges. IEEE Pervasive Comput. **17**(2), 11–21 (2018)
8. Black, M.H., Chen, N.T.M., Iyer, K.K., Lipp, O.V., Bölte, S., Falkmer, M., Tan, T., Girdler, S.: Mechanisms of facial emotion recognition in autism spectrum disorders: insights from eye tracking and electroencephalography. Neurosci. Biobehav. Rev. **80**, 488–515 (2017)
9. Bossavit, B., Parsons, S.: Outcomes for design and learning when teenagers with autism co-design a serious game: a pilot study. J. Comput. Assist. Learn. **34**(3), 293–305 (2018)
10. Boucenna, S., Narzisi, A., Tilmont, E., Muratori, F., Pioggia, G., Cohen, D., Chetouani, M.: Interactive technologies for autistic children: a review. Cogn. Comput. **6**(4), 722–740 (2014)
11. Cai, Y.: Interactive and Digital Media for Education in Virtual Learning Environment. Nova Sciences Publishers, New York (2011)
12. Cai, Y.: 3D Immersive and Interactive Learning. Springer, Singapore (2013)
13. Cai, Y., Goei, S.L.: Simulation Serious Games and Applications. Springer, Singapore (2013)
14. Cai, Y., Chia, N.K.H., Thalmann, D., Kee, N.K.N., Zheng, J., Thalmann, N.M.: Design and development of a virtual dolphinarium for autistic children. IEEE Trans. Neural Syst. Rehabil. Eng. **21**(2), 208–217 (2013)
15. Cai, Y., Goei, S.L., Trooster, W.: Simulation and Serious Games for Education. Springer, Singapore (2016)
16. Caria, S., Paternò, F., Santoro, C., Semucci, V.: The design of web games for helping young high-functioning autistics in learning how to manage money. Mobile Netw. Appl. **23**(8), 1–14 (2018)
17. Caronna, E.B., Milunsky, J.M., Tager-Flusberg, H.: Autism spectrum disorders: clinical and research frontiers. Arch. Dis. Child. **93**(6), 518–523 (2008)
18. Carpenter, L.A., Soorya, L., Halpern, D.: Asperger's syndrome and high-functioning autism. Paediatr. Ann. **38**(1), 30–37 (2009)
19. Coco, M.D., Leo, M., Carcagnì, P., Famà, F., Spadaro, L., Ruta, L., Pioggia, G., Distante, C.: Study of mechanisms of social interaction stimulation in autism spectrum disorder by assisted humanoid robot. IEEE Trans. Cogn. Dev. Syst. **10**(4), 993–1004 (2018)
20. Craig, F., Margari, F., Legrottaglie, A.R.: A review of executive function deficits in autism spectrum disorder and attention-deficit/hyperactivity disorder. Neuropsychiatr Dis Treatment **12**, 1191–1202 (2016)
21. Densmore, A.E.: Helping children with autism become more social: 76 ways to use narrative play. Praeger Publishers, Westport, CT (2007)
22. DiPietro, J., Kelemen, A., Liang, Y., Sik-Lanyi, C.: Computer- and robot-assisted therapies to aid social and intellectual functioning of children with autism spectrum disorder. Medicina 54 (440) (2019)
23. Grandin, T., Scariano, M.M.: Emergence: labeled autistic. Novato, CA (1986)
24. Grossard, C., Hun, S., Dapogny, A., Juillet, E., Hamel, F., Jean-Marie, H., Bourgeois, J., Pellerin, H., Foulon, P., Serret, S., Grynszpan, O., Bailly, K., Cohen, D.: Teaching facial expression production in autism: the serious game JEMImE. Creat. Educ. **10**, 2347–2366 (2019)

25. Grossard, C., Palestra, G., Xavier, J., Chetouani, M., Grynszpan, O., Cohen, D.: ICT and autism care: state of the art. Curr. Opin. Psychiatry **31**(6), 474–483 (2018)
26. Huijnen, C., Lexis, M., Jansens, R., Witte, L.D.: Roles, strengths and challenges of using robots in interventions for children with autism spectrum disorder (ASD). J. Autism Dev. Disord. **49**(2), 1–11 (2019)
27. JISC: Game-based learning. Available online https://www.jisc.ac.uk/media/documents/publications/gamingreportbp.pdf (2013)
28. Khabbaz, A.H., Pouyan, A.A., Fateh, M., Abolghasemi, V.: An adaptive learning game for autistic children using reinforcement learning and fuzzy logic. J. AI Data Min. **7**(2), 321–329 (2019)
29. Khowaja, K., Salim, S.S.: Serious game for children with autism to learn vocabulary: an experimental evaluation. Int. J. Human-Comput. Interact. **35**, 1–26 (2018)
30. Kim, S.H., Bal, V.H., Lord, C.: Longitudinal follow-up of academic achievement in children with autism from age 2 to 18. J. Child. Psychol. Psychiatry **59**(3), 258–267 (2018)
31. Liu, G.-Z., Wu, N.-W., Chen, Y.-W.: Identifying emerging trends for implementing learning technology in special education: a state-of-the-art review of selected articles published in 2008–2012. Res. Dev. Disabil. **34**(10), 3618–4362 (2013)
32. Loh, C.S., Sheng, Y., Ifenthaler, D.: Serious Games Analytics—Methodologies for Performance Measurement, Assessment, and Improvement. International Publishing, Springer (2015)
33. Malaia, E., Cockerham, D., Rublein, K.: Visual integration of fear and anger emotional cues by children on theautism spectrum and neurotypical peers: an EEG study. Neuropsychologia **126**, 138–146 (2019)
34. Marchi, E., Schuller, B., Baird, A., Baron-Cohen, S., Lassalle, A., O'Rielly, H., Pigat, D., Robinson, P., Davies, I., Baltrusaitis, T., Golan, O., Fridenson-Hayo, S., Tal, S., Newman, S., Meir-Goren, N., Camurri, A., Piana, S., Bolte, S., Sezgin, M., Alyuz, N., Rynkiewicz, A., Baranger, A.: The ASC-inclusion perceptual serious gaming platform for autistic children. IEEE Trans. Comput. Intell. AI Games 1–12 (2018)
35. Mehmood, F., Ayaz, Y., Ali, S., Amadeu, R.D.C., Sadia, H.: Dominance in visual space of ASD children using multi-robot joint attention integrated distributed imitation system. IEEE Access **7**, 168815–168827 (2019)
36. Munoz, R., Villarroel, R., Barcelos, T.S., Riquelme, F., Quezada, A., Bustos-Valenzuela, P.: Developing computational thinking skills in adolescents with autism spectrum disorder through digital game programming. IEEE Access **6**, 63880–63889 (2018)
37. Myers, S.M., Johnson, C.P., The Council on Children with Disabilities.: Management of children with autism spectrum disorders. Pediatrics **120**(5), 1162–1182 (2007)
38. Oliveira, M., Guebert, M., Nohama, P.: TEACCH methodology-based web system to support learning for children with autism. IEEE Lat. Am. Trans. **16**(11), 2698–2705 (2018)
39. Organization for Autism Research (OAR) Inc.: Life Journey Through Autism: Navigating the Special Education System, The guidebook. Available online: https://files.eric.ed.gov/fulltext/ED542647.pdf (2012)
40. Silva, S.D., Neto, F.M.M., Lima, R.M.D., Silva, P.D.A., Demoly, K.R.D.A., Cruz, I.N.D.: A serious game as an auxiliary tool for the learning process of children with ASD. In: Handbook of Research on Immersive Digital Games in Educational Environments, pp. 524-553 (2019). ISSN: 2326-8905
41. Simões, M., Bernardes, M., Barros, F., Castelo-Branco, M.: Virtual travel training for autism spectrum disorder: proof-of-concept interventional study. JMIR Serious Games **6**(1) (2018)
42. Tang, J.S.Y., Falkmer, M., Chen, N.T.M., Bölte, S., Girdler, S.: Designing a serious game for youth with ASD: perspectives from end users and professionals. J. Autism Dev. Disord. **49**, 978–995 (2019)
43. Tsikinas, S., Xinogalos, S.: Studying the effects of computer serious games on people with intellectual disabilities or autism spectrum disorder: a systematic literature review. J. Comput. Assist. Learn. **35**, 61–73 (2018)

44. Wong, V.C.N., Hui, S.L.: Epidemiological study of autism spectrum disorder in China. J. Child Neurol. **23**(1), 67–72 (2008)
45. Zhang, Y., Song, W., Tan, Z., Zhu, H., Wang, Y., Lam, C.M., Weng, Y., Hoi, S.P., Lu, H., Chan, B.S., Chen, J., Yi, L.: Could social robots facilitate children with autism spectrum disorders in learning distrust and deception. Comput. Hum. Behav. **98**, 140–149 (2019)
46. Zheng, Z., Zhao, H., Swanson, A.R., Weitlauf, A.S., Warren, Z.E., Sarkar, N.: Design, development, and evaluation of a noninvasive autonomous robot-mediated joint attention intervention system for young children with ASD. IEEE Trans. Human-Mach. Syst. **48**(2), 125–135 (2018)

An Investigation of South African Pre-service Teachers' Use of Simulations in Virtual Physical Sciences Learning: Process, Attitudes and Reflections

Umesh Ramnarain and Mafor Penn

Abstract The use of virtual learning technologies has become of great interest in twenty-first century science education research. Simulations and other virtual learning technologies have been shown to enhance conceptual understandings of abstract scientific concepts and foster motivation and interest in science learning. This book chapter reports on research of South African pre-service teachers' use of simulations in physical sciences learning. In adopting an explanatory sequential mixed method approach, we investigated the experiences of fifty ($n = 50$). Pre-service physical science teachers in the use of PhET simulations explored instructional scaffolding within simulations that supported learner engagement. Further to this, we report on attitudinal changes of these pre-service teachers towards the subject, before and after virtual learning interventions. Findings from investigations suggest that learning by simulations in virtual spaces enhances attitudes towards physical sciences, with post-test attitude scores being significantly higher than pre-test attitude scores. From qualitative data sets, pre-service teachers asserted that their visualization of microscientific phenomena was enhanced. Evidence also suggested that self-directed learning was promoted by the use of simulations. From the research, themes such as convenience of use, interest in science learning, and enablement of guided inquiry emerged. However, pre-service teachers did express concern that simulations lacked authenticity, by failing to replicate hands-on laboratory experiences that enhance the development of manipulative science process skills associated with practical work. Drawing from this research and other studies in this domain, we reflect on the role of virtual learning technologies like simulations in the learning process. We also provide

U. Ramnarain (✉)
Department Science and Technology Education, Faculty of Education,
University of Johannesburg, Johannesburg, South Africa
e-mail: uramnarain@uj.ac.za

M. Penn
Department of Childhood Education, Faculty of Education,
University of Johannesburg, Johannesburg, South Africa
e-mail: mpenn@uj.ac.za

Y. Cai et al. (eds.), *Virtual and Augmented Reality, Simulation and Serious Games for Education*, Gaming Media and Social Effects,
https://doi.org/10.1007/978-981-16-1361-6_7

instructional recommendations for science educators and designers of virtual learning material.

Keywords Attitudes · Scaffolding · Science learning · Simulations · Pre-service science teachers · Virtual laboratories

1 Introduction and Background

In the fourth industrial revolution, technology-enhanced science learning is flagged as a key goal in South Africa's education system [13]. In 2004, the South African Department of Education (DoE) proposed a whitepaper for e-education which advocated the systematic and meaningful integration of learning technologies in teaching and learning [12]. Relevant to global technological transformations, the department anticipated that these learning technologies would act as a motivational tool for learners and enhance critical thinking skills needed for twenty-first century societies. Furthermore, in recent times several simulation-based learning applications have been proposed as alternative pedagogical tools for enhancing students' interest, motivation and visualization of abstract scientific concepts [16, 44]. In physical sciences, laboratory experiences are often regarded as a means by which students' visualization of abstract concepts may be enhanced, leading to science concept formation [29, 31]. However, the overwhelming problem experienced in schools within the South African context is that, traditional physics and chemistry laboratories which facilitate the teaching and learning of physical sciences are a scarce resource [34]. For schools where traditional laboratories are available, students as well as teachers tend to be sceptical about the use of these laboratories in relation to learning science concepts [47]. Some of the reasons advanced for this position are that, laboratory processes tend to be incomprehensible, apparatus are difficult to use or simply not available and the instructional guidance provided is not usually enough to scaffold the understanding of certain concepts [38].

Scaffolding entails the ability to progressively move students from a place of little or no understanding to one of greater understanding and eventually more autonomy in the learning process [49]. Like physical scaffolding, traditional scaffolding in learning describes the teacher's ability to provide progressive levels of support on which a student attains higher levels of skills, knowledge and attitudes [7, 39] in a particular subject. In virtual and e-learning this kind of scaffolding is embedded in the learning design [33] of particular simulations or virtual laboratories. When considering virtual science instruction especially, the scaffold prompts are systematically embedded to navigate students from the least to the most difficult simulations in order to minimize cognitive overload [8], but also to ensure that self-directed learning is enhanced.

For the studies reported in this chapter, we identified a twofold problem which entailed firstly investigating how learning simulations relate to affective constructs like attitude towards physical sciences and secondly acknowledging the gap in research, especially within the South African context on scaffolded instruction of learning in a virtual learning environments. Underlying this study is a national strategy towards equipping under-resourced schools with e-learning resources (smartboards, tablets, computers, etc.). So that the quality of science teaching and learning may be enhanced. This development prioritizes the need for pre-service teachers to be adequately prepared to harness the potential of such e-learning resources when they take up appointments at these schools. Accordingly, this chapter reports on research of third year pre-service teachers' (herein also referred to as students). Attitudinal changes towards physical sciences learning post-learning interventions with simulations. We further assess the role of scaffolding in the virtual learning process. The following research questions guided our investigation:

How do pre-service teachers' attitudes towards physical sciences learning differ pre- and post-learning interventions with virtual learning simulations?

What is the role of implicitly embedded instructional scaffolding in physical sciences learning simulations?

1.1 Aims and Objectives

The main aim of the study was to assess pre-service science teachers' attitudes towards physical sciences learning pre- and post-intervention with virtual laboratory simulations. The study further examined the role of instructional scaffolding implicitly embedded within the virtual learning simulations. To achieve this aim, the following objectives were set.

- To assess pre-service teachers' attitudes towards physical sciences learning prior to learning intervention with virtual laboratory simulations.
- To provide learning interventions using PhET and other simulations for the identified physics and chemistry concepts (Faraday's law and stoichiometry).
- To re-assess pre-service teachers' attitudes post-intervention with virtual laboratory simulations.
- To assess the role of implicitly embedded instructional scaffolding within learning simulations.

1.2 Learning with Simulations

Simulations are defined as imitations of systems and processes, especially for the purpose of studying or showing how things work [21, 44]. Virtual simulation laboratories that are designed by PhET and other virtual laboratories like the ck-12 exploration series provide an excellent platform for the learning of physical science (physics and chemistry) concepts with two-dimensional representations. Prominent design elements of these simulations which distinguish them from diagrammatic representations in textbooks are dynamism and interactivity [23]. Simulations are able to foster interactive and visual learning of microscientific concepts [23, 28, 38]. Even though simulations as used on desktop computers and tablets may not have the affordances of highly immersive three-dimensional virtual reality and other serious games, they do provide the opportunity for learners to engage interactively with simulated objects (simulators), including field lines, forces, atoms, electrons, compounds and elements when learning science concepts. In fact, Rutten et al. [42] also suggest that when compared to textbooks and traditional lectures learning with simulations provides students with a platform to interact with phenomena, practice tasks and explore hypothetical situations.

The PhET simulation laboratories originally developed by the University of Colorado provide students with a platform to experiment and manipulate different variables that would not be possible when using traditional learning strategies [10, 32, 42]. One of the prominent features of learning simulations is the embedded scaffolding of activities such that students are able to migrate from easier to more difficult concepts. Some learning simulations are also furnished with self-assessment tasks, which aid students to assess their understandings of concepts and enhance collaboration with their peers [23]. These together with other benefits of virtual learning such as the global access to learning resources, multiple representations, pacing and self-directed autonomous learning are some of the highlights that were exploited in a five week learning intervention with participant pre-service teachers.

1.3 Attitude

Attitude is an affective construct which refers to a mental and emotional entity characterized by a person's actions or thoughts towards an object or subject [36]. This construct has been extensively researched in science learning [22] due to the strong correlation between attitude and other constructs like science learning, achievement in science and even orientation towards careers in science [5, 9, 35]. For the study reported in this chapter, we premised that if pre-service teachers showed a positive attitude towards learning physics concept with virtual simulated laboratories, it will facilitate their conceptual understanding of physics concepts and their teaching practice when they eventual became in-service teachers. Though the main focus of this study was on attitudinal outcomes, it was also deemed relevant to investigate the instructional processes within the actual simulations as students engage with them in learning physical sciences concepts.

1.4 *Instructional Scaffolding in Virtual Learning Simulations*

PhET simulations developed by the University of Colorado Boulder are free virtual learning tools, which incorporate implicit scaffolding whereby concepts could be learned with minimal explicit guidance from a teacher [33, 37]. Implicit scaffolding does not include verbal or written instruction but is embedded systematically in the design of a virtual laboratory simulations or illustrations. The implicitly scaffolded design features of a simulation facilitate the learning process such that the student moves through the zone of proximal development (ZPD), without much guidance from an instructor or laboratory manual [2]. This kind of instructional scaffolding is not entirely new as some researchers postulate that there are "three roles software could play to provide scaffolding: communicating processes to learners, coaching learners with hints and reminders about their work, and eliciting articulation from learners to encourage reflection" [39, p. 338]. Scaffolding also provides a platform where students could attain individualized learning outcomes without interfering with collective learning paces [39, 40]. Virtually embedded implicit scaffolding differs from the explicit traditional scaffolding in that when embedded in simulations, the scaffolding tool not only provides necessary guidance to students but changes the nature of students' perceptions and reception of the guidance [24]. That is, students feel they are in control of their learning and do not feel as instructed as in the case of explicit scaffolding.

2 Theoretical Framework

Active engagement and student involvement in gaining conceptual understandings are an integral aspect of physical sciences learning due to the abstract nature of certain concepts. An inquiry-based hands-on pedagogical approach, which motivates students to actively engage with concepts rather than receive passive instruction, aids to enhance visualization of abstract concepts [20]. Underpinning the research is a set of learning theories which are related to the ways in which technology, multiple media and experimental learning all contribute in fostering learner-centred learning of science concepts and eventually more positive attitudes towards a subject [38].

Firstly, as postulated by early constructivist theorists, experiential and experimental learning have the ability to enhance cognition significantly [3]. By implication, it is necessary for learners to be actively involved in investigating concepts, through the use of experiments in which they are active participants and not passive recipients of knowledge [4, 17]. This kind of active engagement can be attained in traditional science laboratories, virtual laboratories, fieldwork and broad scientific investigations. Laboratory-based science teaching and learning has been shown to play a significant role in learners' understandings of abstract scientific concepts

[14]. Cognitive constructivists also advocate that, in designing learning activities instructors should be considerate of the learner's context, as daily experiences and context play a huge part in how learners learn. In the twenty-first century, learning with technology constitutes a major part of learner attribute and should be incorporated in the learning of science [20, 25, 45]. One would naturally wonder how gains in conceptual understandings relate to learners' attitude. Several studies [5, 9, 35] have shown that when students acquire an understanding of concepts taught to them, they tend to exhibit positive attitudes towards the subject where concepts are learned.

A second theory relates to the instructional scaffolding implicitly embedded in the design of the learning simulations. Podolefsky et al. [37] proposed an emerging theoretical framework for implicit scaffolding stemming from the Vygotskian constructivist theories. The Vygotskian theories portray scaffolding as the systematic and stepwise support for transitioning a student through the zone of proximal development (ZPD), from what they cannot learn on their own, to the skills, knowledge and attitudes they can learn autonomously [6]. In science subjects like physics and chemistry which constitute the South African physical sciences curriculum, scaffolding is essential in navigating students through abstract scientific concepts which maybe otherwise difficult to grasp with unmediated direct instruction. Normally scaffolding is explicit whereby an instructor provided verbal instructions, systematic probing via questions and puzzles or even demonstrations in order to provoke students' critical thinking. These kinds of explicit scaffolding tend to be very teacher-centred and though helpful in achieving learning goals, decreases students' critical thinking abilities. The nature of implicit scaffolding is such that students are guided without necessarily feeling guided [2] and can navigate their learning in a virtual space without extensive tutor assistance [6].

3 Identified Physical Sciences Concepts

In the empirical study, participant pre-service teachers participated in an anonymous survey to select difficult physical science concepts that were most challenging in their second year physical sciences module. Faraday's law of electromagnetic induction (EMI) and Newton's third law on the conservation of momentum in relation to collisions and velocity were identified as the most abstract and difficult physics concepts in the course. For the chemistry component, stoichiometry, aspects or organic chemistry and generally chemical reactions topped the difficult chart.

Pre-service teachers indicated that learning Faraday's law of electromagnetic induction was difficult because, the relationship between electricity and magnetism was complex due to a lack of visualization. They did not comprehend the time changes (dt) in a magnetic field and the several challenges associated to the flux rule are also reported in studies such as Zuza et al. [51] and Jelicic et al. [19]. A learning intervention was then designed to address difficulties that students encounter when learning concepts on electromagnetic induction and stoichiometry using available PhET laboratory simulations and the embedded activities in the simulated learning environment. Figure 1 below shows a screenshot of the initial PhET laboratory that was used in exploiting the relationship between flux and current as suggested by Faraday's law.

The screenshot in Fig. 1 depicts a phase where pre-service teachers were able to manipulate the flux across the different number of turns of a coil, measuring the electromagnetic force (EMF). As they changed other variable such as time, direction and magnetic poles, the students followed stepwise instructional guidance within the simulator, starting with a few coil turns and systematically increasing number of turns and recording changes in flux. After this phase, they advanced to Faraday's electromagnetic laboratory in further investigating the applications of Faraday's law and the possible relationships between flux and current in generators and transformers as seen on the screenshot in Fig. 2.

The screenshot in Fig. 2 shows a multiscaffolded simulation "*tabs*" where students interact with the content from a bar magnet and its properties to a generator, altering the magnetic poles and directions of field lines while measuring the effects on the flow of induced current as stipulated by Lenz's law (induced current opposes changed magnetic fields).

In chemistry, the application of stoichiometry in the balancing of chemical equations was flagged by third year students as being difficult. Findings from some studies also support that these perceived difficulties with stoichiometry are

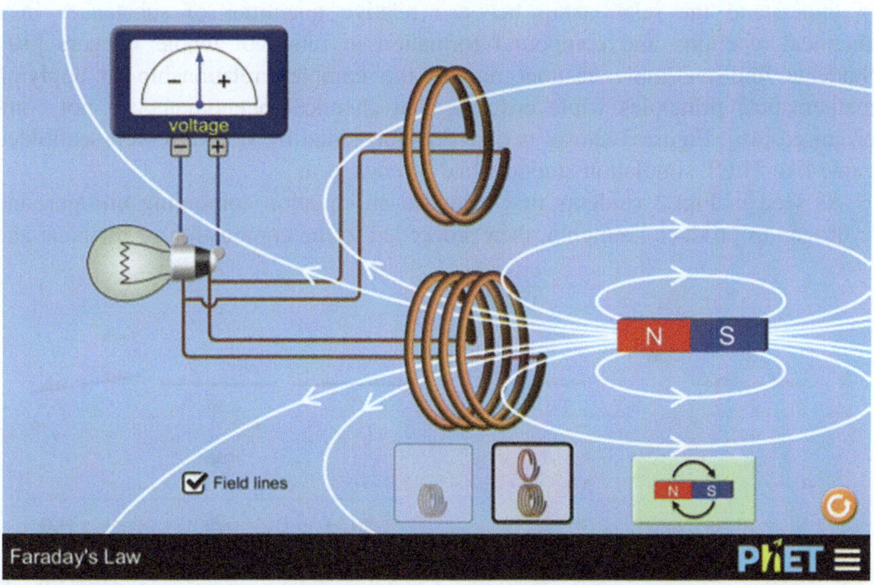

Fig. 1 Screenshot of Faraday's law. *Source* https://phet.colorado.edu/en/simulations/faraday's-law

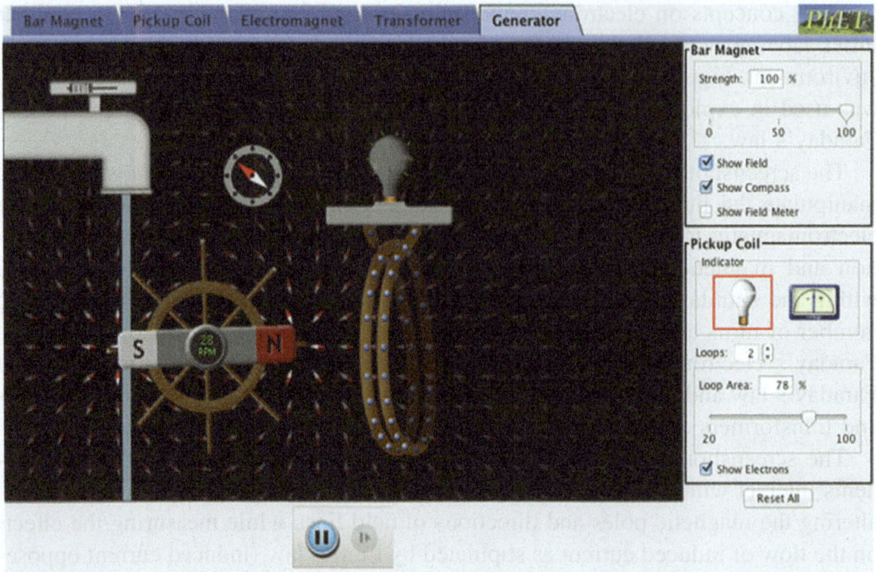

Fig. 2 Screenshot applications of Faraday's law. *Source* https://phet.colorado.edu/en/simulation/generator

associated with the manipulation of numbers which constitute the direct relationship between mathematics and chemistry [30]. In learning stoichiometry, students need to understand the relationship between relative quantities of substances in a chemical reactions and compound formation in ratios of whole integers [30]. Students find the topic abstract due to the complex relationship in applying mathematical principles while ensuring that chemical phenomena are not compromised [46]. Figure 3 shows two screenshots sided by side of a well-scaffolded game-like PhET simulation students had to engage in.

As seen in Fig. 3 students first balanced an equation combining nitrogen and hydrogen to produce ammonia, then proceeded to the combustion of methane and

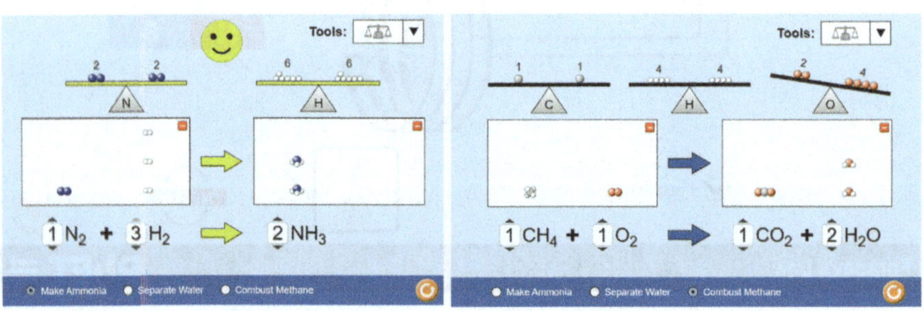

Fig. 3 Screenshot showing levels of scaffolding in balancing chemical equations. *Source* https://phet.colorado.edu/sims/html/balancing-chemical-equations/latest/balancing-chemical-equations_en.html

more complex reactions where their knowledge of aspects of stoichiometry was tested. This particular simulation has a gamification feel to it as rewards are embedded in the form of "smileys" when an equation is adequately balanced. A scale tool is also included to indicate imbalance where a particular element in the equation has not been balanced. With this kind of embedded instant feedback system, *students may be able to self-assess*, self-correct and eventually proceed to tackling the balancing of more complex chemical equations.

4 Method

An explanatory sequential mixed methods design [11] was employed whereby, a quantitative survey was firstly used to assess pre-service physical sciences teachers' attitudes before and after learning interventions with PhET simulation laboratories. This phase was then followed by a qualitative inquiry about the role of instructional scaffolding embedded within simulations and student reflections on their simulation-based learning experience. In enacting the quantitative inquiry, a quasi-experimental one-group pre-test–post-test design was preferred as it is well suited within educational contexts. Fifty (n = 50) pre-service physical sciences were purposively selected at a South African tertiary institution. The criterion for selecting teachers included major subject preferences and the subject pre-service teachers intended to teach once they had acquired their Bachelor of Education qualification. 22 (44%) of the participants were females and 28 (56%). were males, with participants' ages ranging between 20 and 25 years old. 86% of participants belonged to the age group 20 and 21 year old.

The validated Physics Attitude Scale (PAS) was adapted from Kaur and Zhoa [22]. It was used for data collection in the first phase of the study. The adapted instrument consists of 22 items designed with guidelines, which describe two attitudinal constructs relating to students' enthusiasm towards physics and how they learn and understand the subject. In adapting the PAS, the researchers modified the wording of the items to fit the context of simulations without altering the constructs therein. A 5-point Likert scale (Likert 1932) was used for the adapted PAS responses ranging from Strongly Agree-5, Agree-4, Neutral-3, Disagree-2, to Strongly Disagree-1. The items on the adapted PAS were easy to score and had robust psychometric properties including good internal consistency for items in each construct, content validity by experts and good readability for participants (see full validation in [22]. Negative items were also reverse scored during analysis. SPSS 26 was used to analyse the internal consistency of items within the constructs of the adapted instrument. The instrument was administered to all participants before the commencement of learning interventions. Internal reliability was calculated and inter-item reliability for the constructs showed that the attitude scale was of good fit with Cronbach's alpha $\alpha \geq 0.88$. The adapted PAS was again administered to participants at the end of a five weeks virtual learning intervention using simulations and post-test scores analysed.

To ensure interactive learning, participants of the study were carefully assigned to one of 10 groups, with each group comprising of five students to engage with simulations during the learning interventions. In each group, students at all ability levels based on achievement were represented. The pre- and post-intervention adapted PAS scores were analysed using descriptive and inferential statistics, to show means, standard deviations and paired samples t-test, respectively.

For the qualitative part of the research, 10 focus group semi-structured interviews were conducted with all participants of the study, to elaborate on their learning experiences with simulations as a whole and specifically their reflections on the role of scaffolded instructions embedded within the simulation interface. Interviews were audio-recorded and data transcribed verbatim to textual data for ease of analysis. ATLAS.ti version 8, a computer-assisted qualitative data analysis software (CAQDAS), was also used to facilitate the content analysis of transcribed textual data. Table 1 shows a sample of student responses to the interview questions.

Based on the participant pre-service teacher responses we then coded data and further categorized the common codes [43], to generate the themes discussed as part of the findings.

5 Results and Discussions

5.1 Physics Attitude Scale (PAS). Scores Pre- and Post-intervention

This section examines the results and discussions of the adapted Physics Attitude Scale (PAS) cores before and after the learning interventions. Table 2 shows a sample of pre-service teachers' mean scores and standard deviations for some adapted PAS items pre- and post-interventions with simulation-based virtual learning.

As reflected in the extract in Table 2 pre-service teachers' mean PAS scores post-intervention for 21 items were higher than the pre-intervention PAS scores for the same items. Only item number 16, *demonstrating physical sciences concepts in class, is not time consuming* recorded the same Pre- and Post-intervention PAS scores.

A paired sample *t*-test was conducted to establish if the observed mean differences between pre- and post-test were significant. The data was checked for normality and confirmed to be normally distributed. The result from *t*-test after normality had been established confirmed that, there was a significant difference between Pre- and Post-intervention PAS scores as seen on Table 3.

As seen on Table 3 pre-service teachers' attitude scores post-intervention with simulation-based learning interventions were significantly higher ($M = 84.04$, SD = 5.345), than the attitude scores pre-intervention ($M = 58.42$, SD = 14.444),

Table 1 Sample student responses to interview questions

Questions	Group responses
How did you find the PhET simulation laboratories and the activities? Provide a brief explanation	Grp1: We found the simulations to be very engaging and interactive Grp3: The simulation on balancing equations was very entertaining and felt like a game Grp 6: It was easy to navigate the simulation interface as they had some kind of hidden prompts
In your own opinion what was the role of the embedded prompts within the simulation interface? Please elaborate on your reflection	Grp 4: Simulation user interface had different tabs which navigated us stepwise towards the different learning objectives that were provided to us by the lecturer Grp 5: Particularly with stoichiometry/ balancing equations, we enjoyed how a smiley and a scale balance tool were able to guide if we were correct or wrong. Wow! So fascinating. When we made an error with the equations, the scale balanced was uneven and we tried again until the product–reactant balance was attained and we got that smiley face. Overall, it felt great to be in charge of our own learning Grp 1: In Faraday's lab simulations we were able to manipulate variable answering our own research question within the simulation
After the intervention classes, did you need help with the activities and concepts learned in with simulations? Explain your reasoning	Grp 7: We could easily work on my own and self-pace. We mostly did not require help form the tutor as we used to do previously Grp 10: We were able to identify and correct our mistakes as a group without much assistance. The prompts in the simulation interface were really good for us Grp 9: It was easy to remediate and adequately pace if you were falling behind without assistance
Did the PhET simulations improve your conceptual understandings/ attitudes towards the learned physics concept in any way?	Grp 8: Yes, we could visualise, field lines, electrons, movements, etc., something that was not easy when we use textbooks and even traditional lab. This made understanding easy Grp 5: Our whole perspective on physical sciences course has been changed. Most of us in the group now look forward to learning new physical sciences using different simulations Grp 3: Yes very much, because we could try different experiments that enabled us to critically analyse some of the concepts in EMI. Our performance improved as a group and individually

(continued)

Table 1 (continued)

Questions	Group responses
	Grp 1: For me yes because it was easy. Yes our group had so many improvements as we were able to tackle different difficulty level activities over and over until we understood
Sum up your overall experience with reference to gains and possible disadvantages of simulations in physical sciences learning	Grp 9: We were intrigued by the learning possibilities with simulations as a whole and hoped to use them in our teaching practice Grp 10: On a positive note we were able to repeat tasks over and over; Grp 2: We kept practicing our concepts Grp 6: We recorded so many learning gains and our attitudes towards physical sciences has really changed for the better Grp 7: The simulations were, interactive, fun, engaging and informative in ways that were fascinating to us Grp 1: With the PhET simulations the good thing is that if you don't get it then you do it over and over again Grp 2: On the negative side we found could easily create some misconceptions Grp 4: Misconceptions can be promoted and one needs to be careful when using simulations Grp 5: The screen exposure was a little worrying when you think of young learners. However, science process skills are not learned by the click of a mouse Grp 3: We noted that though learning was promoted in the simulation labs, this could not replace traditional science lab especially when we refer to the acquisition of science process skills Grp 6: Simulations are in a machine, which is just…not tangible when I think of it. Authentic learning in real spaces will always be needed for science

$t(49)$. = 17.429, $p < 0.01$. This showed a positive change in attitude towards learning physical sciences with the introduction of simulation. The qualitative phase of the study that is now presented provided some insights to this observed trends.

Table 2 Pre- and post-intervention PAS scores

PAS items	Post-test		Pre-test	
	Mean	SD	Mean	SD
1. Learning physical science phenomena and their description is most enjoyable to me	3.96	0.81	2.62	1.03
2. Studying topics in physical sciences in detail is worth it	4.24	0.63	2.96	0.99
3. My confidence level increases when doing physical sciences investigations in simulated laboratory environment	4.22	0.68	2.54	1.18
4. The basic knowledge of physical science is useful for everyone	4.06	0.51	3.06	0.98
5. Physical science is not a boring subject for me	4.50	0.65	2.62	1.24
6. The successful completion of a physical sciences investigation excites me to do other experiments	4.34	0.56	2.48	1.11
7. I will be happy if the practical work in physical sciences is increased so that I may devote less time in studying theory	4.38	0.57	2.64	1.05
8. I am punctual with physical sciences learning tasks	3.74	0.80	2.40	1.13
9. I wait eagerly for physical sciences simulation-based tutorials	3.86	1.05	2.74	1.10

Table 3 Paired sample *t*-test for the adapted PAS

Pair 1	*n*	Mean	SD	Std. error mean	*t*	*p*
Pre-PAS scores	50	58.42	5.35	0.76	17.43	$p < 0.01$
Post-PAS scores	50	84.04	14.44	2.04		

5.2 *Findings from Semi-structured Interviews*

Follow-up semi-structured focus group interviews conducted with all participants revealed that the students found the PhET simulations quite useful in enhancing their learning experiences and their attitudes towards the learning of physical sciences. Based on a content analysis of data from the interview phase, students expressed several benefits on how simulations influenced their learning and particularly their attitudes towards physical sciences learning. With a focus on the research question "*What is the role of implicitly embedded instructional scaffolding in physical sciences learning simulations?*" pre-service teachers revealed that, embedded guidance within simulations was especially helpful in pacing their learning and enhancing autonomous learning. Other themes that emanated included, the promotion of guided and systematic inquiry, learning convenience, interest in science learning, repetitive remedial learning and adequate cognitive scaffolding. Some pitfalls of simulations described by students included a lack tangibility and authenticity especially with regard to the development of science process skill.

In the sections below, we provide explanations of the assertions that related to the nature of scaffolded instruction within simulations.

5.2.1 Self-Directed and Autonomous Learning

Nine of the ten groups of participants indicated that, the main role of the scaffolded instructions within simulations was the ease with which students were able navigate through the learning tasks without guidance from the tutor or lecturer. They further explained that for certain tasks there were three interfaces starting from and easier to a more difficult and complex concept at the third interface. Students also mentioned that because feedback was instant and they could manipulate variables within simulations, it became easier to self-teach and self-assess one's ability and understandings. For the stoichiometry tasks, a balancing scale was embedded within the simulations to indicate where students missed balancing an element. This afforded them an opportunity to correct their effort, without them being told the correct answer, such as would be the case when they are instructed by a lecturer. With implicitly embedded scaffolds, students found learning quite fascinating and felt they could navigate their way through the simulations to attain learning outcomes. This had a positive impact on their attitudes and confidence level in handling physical sciences learning tasks.

5.2.2 Systematic Guided Inquiry

All 10 groups of students indicated that, scaffolded instructional guidance made it easy for inquiry-based learning to be enacted systematically within simulations. They were awed by how easy it was to pose a scientific question about the relationship between two variables like flux and current, then manipulate the variables within simulations and collect relevant data to answer the posed question. This affordance was fascinating to students as they observed that there was again no need for much instructor guidance.

5.2.3 Cognitive Scaffolding

More than 80% of the participants also indicated that scaffolded instructions within simulations aided their retention of learned concepts and made it easy for them to assimilate concepts. They also emphasized that stepwise guidance reduced their cognitive load, as they were not overwhelmed with the learning of many concepts at a time. They indicated that as with physical scaffolding, concepts were layered stepwise from what was already known as prior knowledge to the new concepts unknown to them.

5.2.4 Remediation

One of the affordances of implicitly scaffolded instruction was the reset key embedded in the simulation interface. Participant students found this rather useful for repeating entire simulations or sections where they did not understand or needed second chances.

5.2.5 Moderation of Learning Pace

Participant students recognized the role that scaffolded instruction played in ensuring that different groups attained learning goals according to their learning abilities. For example, within the stoichiometry simulation, pace was well negotiated based on the students' learning abilities such that no one was under any pressure to catch up to those who were faster to grasp concepts or wait in order to accommodate those who were lagging. Everyone systematically made progress as they attained learning goals and milestones.

6 Discussion

The findings reported in this chapter revealed that learning through the use of simulations had positive effects on pre-service teachers' attitudes towards physical sciences learning. These attitudinal changes were further supported by the findings from semi-structured interviews which provided complementary evidence for the changes and perceived learning gains. In relation to literature, the findings from the study corroborate with findings from other recent studies [9, 14, 17, 32] where using different simulation-based learning strategies saw students recording positive attitudes and improved achievement. In the above-mentioned studies, virtual laboratory learning by immersive or non-immersive simulations was reported to have a positive impact on learners' attitudes and motivation towards science learning in general. Arvind and Heard [1] also reported that, the use of virtual laboratories simplified complex physics concepts and changed students' negative perceptions of a physics course. Similarly, Tuysuz [48] found that, students who were comfortable in using virtual simulation laboratories showed a more positive attitude towards learning chemistry concepts. However, contrary findings also exist, with Faour and Ayoubi [15] reporting no attitude differences in a study, which assessed grade 10 students' attitudes towards physics post-intervention with virtual laboratory simulations.

From the semi-structured interview phase, pre-service teachers indicated that they derived several benefits from using the virtual simulation laboratories for physical sciences learning. Gains included the fundamental role that scaffolded instruction had in simplifying the learning tasks within simulations, anytime learning convenience, increased confidence in handling physical sciences problems, more autonomy in doing learning tasks and the general feeling of enjoyment and play associated with the simulations. Similar findings were accrued in a study by Moore et al. [33], who revealed that the nature of implicit scaffolding embedded in simulations laboratories was able to guide students in attaining learning outcomes with minimal explicit instructions from instructors and tutors. Other researchers such as Chamberlain et al. [6], and Chen and Law [8] also found that scaffolded guidance enhanced time on task engagement with concepts under investigation. Theoretically, the scaffolded design embedded in simulations also enhances guided inquiry [40]. Reeves and Reeves [41] further reported improvements in self-directed learning with immediate feedback on tasks.

7 Conclusions

The findings from this study have implications for simulation-based learning in science. Based on the findings, we conclude that the use of simulations provides a platform for participants to experiment with and visualize abstract physical sciences concepts, in a properly scaffolded simulation design. It was worth observing that the pre-service teachers felt that virtual learning simulations could not replace learning via experimentation in traditional science laboratories. However, simulations are capable of complementing physical sciences learning experiences in yielding positive attitudes and cognitive gains. We also conclude that, the implicitly scaffolded instructional guidance within simulations provide leverage for students to learn and interact with the simulator better in a virtual space and suggest that further studies be exploited on the different levels of embedded scaffolding for concept-specific learning processes, in virtual-science learning. Though students indicated that they felt the acquisition of science process skills could be hampered when using simulations, we postulate that using simulation could be the first stage of students familiarizing themselves with manipulative science process skills. This ideal is elaborated for instance in studies like Jaakkola et al.[18], who found that students could understand concepts in electricity better when combining the use of simulation and the real circuits in parallel than using only computer simulations. Other studies also advocate for the complementary use of simulations in combination with traditional laboratory settings [29, 50].

Based on these conclusions, we recommend that, curriculum experts situate simulations in science curricula globally, for complementing traditional science learning and enhancing inquiry-based learning. For science education researchers in particular we suggest further large-scale longitudinal studies be undertaken to assess the effectiveness of simulations in science achievement and other affective domains like interest in science careers, motivation and perceptions. For instructional designers, we recommend as a fundamental prerequisite for developing software for science learning that there be a systematic alignment of simulations to learning objectives such that instructions are well-scaffolded from easier to more complex concepts within the simulation interfaces. This is because visualization alone is not enough to enhance conceptual understandings, and implicit guidance will aid students to take charge in their own learning when using simulations and minimize teacher-centred instruction in science learning.

References

1. Arvind, V.R., Heard, J.W.: Physics by simulation: teaching circular motion using applets. Lat. Am. J. Phys. Educ. **4**(1), 35–39 (2010)
2. Bonawitz, E., Shafto, P., Gweon, H., Goodman, N.D., Spelke, E., Schulz, L.: The double-edged sword of pedagogy: instruction limits spontaneous exploration and discovery. Cognition **120**, 322–330 (2011). https://doi.org/10.1016/j.cognition.2010.10.001
3. Bruner, J.S.: Acts of Meaning. Harvard University Press, Cambridge, MA (1990)
4. Cai, Y., Ter Tay, C., Keong Ngo, B.: Introduction to 3D immersive and interactive learning. In: Cai, Y. (ed.) 3D Immersive and Interactive Learning, pp. 1–16. Springer, Singapore (2013). https://doi.org/10.1007/978-981-4021-90-6
5. Can, H.B.: Students' attitudes toward school chemistry: the effect of interaction between gender and grade level. Asia-Pacific Forum on Science Learning and Teaching **13**(1), 11–16 (2012)
6. Chamberlain, J.M., Lancaster, K., Parson, R., Perkins, K.K.: How guidance affects student engagement with an interactive simulation. Chem. Educ. Res. Prac. **15**(4), 628–638 (2014). https://doi.org/10.1039/c4rp00009a
7. Chang, H.-Y., Linn, M.C.: Scaffolding learning from molecular visualizations. J. Res. Sci. Teach. **50**(7), 858–886 (2013). https://doi.org/10.1002/tea.21089
8. Chen, C.-H., Law, V.: Scaffolding individual and collaborative game-based learning in learning performance and intrinsic motivation. Comput. Hum. Behav. **55**, 1201–1212 (2016). https://doi.org/10.1016/j.chb.2015.03.010
9. Chua, K.H., Karpudewan, M.: The role of motivation and perceptions about science laboratory environment on lower secondary students' attitude towards science. EdUHK APFSLT. **18**(2) (2017)
10. Clark, B.D., Tanner-Smith, E.E., Killingsworth, S.S.: Digital games, design, and learning: a systematic review and meta-analysis. Rev. Educ. Res. **86**(1), 79–122 (2016). https://doi.org/10.3102/0034654315582065
11. Creswell, J.W., Creswell, J.D.: Research Design: Qualitative, Quantitative, and Mixed Methods, 6th edn. Sage, Thousand Oaks, CA (2018)
12. Department of Education. White Paper on e-Education. Government Gazette. Government Printers, pretoria (2004)
13. Department of Basic Education: Curriculum and assessment policy statement: grades 10–12 physical sciences. Government Printer, Pretoria (2011)
14. Estapa, A., Nadolny, L.: The effect of an augmented reality enhanced mathematics lesson on student achievement and motivation. J. STEM Edu. 16(3) (2015)
15. Faour, M.A., Ayoubi, Z.: The effect of using virtual laboratory on grade 10 students' conceptual understanding and their attitudes towards physics. J. Edu. Sci. Environ. Health (JESEH). **4**(1), 54–68 (2018).https://doi.org/10.21891/jeseh.387482
16. Hallinger, P., Wang, R.: The evolution of simulation-based learning across the disciplines, 1965–2018: a science map of the literature. simulation & gaming (2019). https://doi.org/10.1177/1046878119888246
17. Hsu, Y.S., Lin, Y.-H., Yang, B.: Impact of augmented reality lessons on students' STEM interest. Res. Pract. Technol. Enhanc. Learn. **12**(2), 1–14 (2017). https://doi.org/10.1186/s41039-016-0039-z

18. Jaakkola, T., Nurmi, S., Veermans, K.: A comparison of students' conceptual understanding of electric circuits in simulation only and simulation-laboratory contexts. J. Res. Sci. Teach. **48**(1), 71–93 (2011). https://doi.org/10.1002/tea.20386
19. Jelicic, K., Planinic, M., Planinsic, G.: Analyzing high school students' reasoning about electromagnetic induction. Phys. Rev. Phys. Educ. Res. **13**(1), 010112 (2017)
20. Jian-Hua, S., Hong, L.: Explore the effective use of multimedia technology in college physics teaching. Energy Procedia 17, 1897–1900; 2012 International Conference on Future Electrical Power and Energy Systems (2012). https://doi.org/10.1016/j.egypro.2012.02.329
21. De Jong, T., van Joolingen, W.R.: Scientific discovery learning with computer simulations of conceptual domains. Rev. Educ. Res. **68**(2), 179–201 (1998)
22. Kaur, D., Zhao, Y.: Development of physics attitude scale (PAS), An instrument to measure students' attitudes toward physics (Article). Asia-pacific Educ. Res. **26**(5), 291–304 (2017). https://doi.org/10.1007/s40299-017-0349-y
23. Kluge, A.: Learning science with an interactive simulator: negotiating the practice-theory barrier. Int. J. Sci. Educ. **41**(8), 1071–1095 (2019). https://doi.org/10.1080/09500693.2019.1590881
24. Latour, B.: Reassembling the Social: An Introduction to Actor-Network Theory. Oxford UP, Oxford (2005)
25. Lee, J., Park, H.: A Study on cases for application of flipped learning in K-12 education. J. Digit Convergence. **14**(8), 19–36 (2016)
26. Likert, R. (1932). A technique for measurement of attitudes. *Archives of Psychology, 140,* 5-55.
27. Makransky, G., Terkildsena, T.S., Mayer, R.E.: Adding immersive virtual reality to a science lab simulation causes more presence but less learning (2017). https://doi.org/10.1016/j.learninstruc.2017.12.007
28. Manunure, K., Delserieys, A., Castéra, J.: The effects of combining simulations and laboratory experiments on Zimbabwean students' conceptual understanding of electric circuits. Res. Sci. technol. Educ. **38**(3), 289–307 (2020) https://doi.org/10.1080/02635143.2019.1629407.
29. Marais, F., Combrinck, S.: An approach to dealing with the difficulties undergraduate chemistry students experience with stoichiometry. S. Afr. J. Chem **62**, 88–96 (2009). Retrieved from https://www.scielo.org.za/pdf/sajc/v62/17.pdf
30. McElhaney, K.W., Chang, H., Chiu, J.L., Linn, M.C.: Evidence for effective uses of dynamic visualisations in science curriculum materials. Stud. Sci. Educ. **51**(1), 49–85 (2015)
31. Merchant, Z., Goetz, E.T., Cifuentes, L., Kenney-Kennicutt, W., Davis, T.J.: Effectiveness of virtual reality-based instruction on students' learning outcomes in K-12 and higher education. Comput. Educ. **70**, 29–40 (2014)
32. Moore, E.B., Herzog, T.A., Perkins, K.K.: Interactive simulations as implicit support for guided-inquiry. Chemistry education research and practice **14**, 257–268 (2013)
33. Naidoo, J., Paideya, V.: Exploring the possibility of introducing supplemental instruction at secondary school level. S. Afr. J. Educ. **35**(2), 10–22 (2015). https://doi.org/10.15700/saje.v35n2a1022
34. Otor, E.E., Achor, E.E.: Effect of concept mapping strategy on students' attitude in difficult chemistry concepts. Eur. J. Edu. Sci. **1**(3), 116–124 (2013)
35. Perloff, R.M.: The Dynamics of Persuasion: Communication and Attitudes in the Twenty-First Century. Routledge, London (2016)
36. Podolefsky, N.S., Moore, E.B., Perkins, K.K.: Implicit scaffolding in interactive simulations: design strategies to support multiple educational goals (2014). Retrieved from https://arxiv.org/abs/1306.6544
37. Potkonjak, V., Gardner, M., Callaghan, V., Mattila, P., Guetl, C., Petrovi, V.M., Kosta, J.: Virtual laboratories for education in science, technology, and engineering: a review. Comput. Educ. **95**, 309–327 (2016). https://doi.org/10.1016/j.compedu.2016.02.002

38. Quintana, C., Reiser, B.J., Davis, E.A., Krajcik, J., Fretz, E., Duncan, R.G., et al.: A scaffolding design framework for software to support science inquiry. J. Learn Sci. **13**(3), 337–386 (2004)
39. Quintana, C., Fishman, B.J.: Supporting science learning and teaching with software-based scaffolding. In: Annual Conference of AERA, San Francisco, CA (2006). Retrieved from https://pdfs.semanticscholar.org/6ce3/8816c8cbc7f76ef8934d4cb63abaacf24aa2.pdf
40. Reeves, T.C., Reeves, P.M.: Designing online and blended learning. In: Hunt, L., Chalmers, D. (eds.). University Teaching in Focus: A Learning Centred Approach. Oxford, Routledge (2012)
41. Rutten, N., van Joolingen, W.R., van der Veen, J.T.: The learning effects of computer simulations in science education. Comput. Educ. **58**, 136–153 (2012). https://doi.org/10.1016/j.compedu.2011.07.017
42. Saldaña, J.: The Coding Manual for Qualitative Researchers. Sage, Thousand Oaks, CA (2016)
43. Smetana, L.K., Bell, R.L.: Computer simulations to support science instruction and learning: a critical review of the literature. Int. J. Sci. Educ. **34**(9), 1337–1370 (2012). https://doi.org/10.1080/09500693.2011.605182
44. Swallow, M.J.C., Olofson, M.W.: Contextual understandings in the TPACK framework. J. Res. Technol. Educ. **49**(3–4), 228–244 (2017). https://doi.org/10.1080/15391523.2017.1347537
45. Taskin, V., Bernholt, S.: Students' understanding of chemical formulae: a review of empirical research. Int. J. Sci. Educ. **36**(1), 157–185 (2014). https://doi.org/10.1080/09500693.2012.744492
46. Trundle, K.C., Bell, R.L.: The use of a computer simulation to promote conceptual change: a quasi- experimental study. Comput. Educ. **54**(4), 1078–1088 (2010). https://doi.org/10.1016/j.compedu.2009.10.012
47. Tuysuz, C.: The effect of the virtual lab. on students' achievement and attitude in chemistry. Int. Online J. Educ. Sci. **2**(1), 37–53 (2010)
48. Wood, P., Bruner, J., Ross, G.: The role of tutoring in problem solving. J. Child Dev. Psychiatry **17**, 89–100 (1976)
49. Zacharia, Z.: Comparing and combining real and virtual experimentation: an effort to enhance students' conceptual understanding of electric circuits. J. Comput. Assist. Learn. **23**, 120–132 (2007). https://doi.org/10.1111/j.1365-2729.2006.00215.x
50. Zuza, K., Almudi, J.M., Leniz, A., Guisasola, J.: Addressing students' difficulties with Faraday's law: a guided problem solving approach. Phys. Rev. Special Topics-Phys. Educ. Res. **10**(1), 010122 (2014)

Examining Pre-service Teachers' Capability to Design Inquiry Learning Activity Sequences with Embedded Simulations

Marios Papaevripidou, Nikoletta Xenofontos, Tasos Hovardas, and Zacharias Zacharia

Abstract In this study, we aimed at examining pre-service teachers' capability to design inquiry learning activity sequences (i.e., inquiry learning spaces or ILSs) after receiving proper training. These ILSs were developed in the context of the Go-Lab ecosystem. The Go-Lab ecosystem includes virtual laboratories (simulations; https://www.golabz.eu/labs) and applications (software scaffolds; see https://www.golabz.eu/apps) that can be embedded in learning activity sequences (inquiry learning spaces), which are developed in a web-based learning environment with the Graasp authoring tool (https://graasp.eu/). Teacher training included presentation of the Go-Lab ecosystem, demonstration of the Graasp authoring tool, and inspection and elaboration on a pilot inquiry learning space. As part of the requirements of two undergraduate university courses, pre-service teachers were asked to develop their own ILS in a context aligned with the national curriculum. As part of the final exam for the courses, they were also asked to transform a lesson plan (an activity sequence in the form of a learning activity sheet) into an ILS. Their ILSs were assessed using a list of criteria developed for the purposes of this study. Our results highlight the strengths and weaknesses of the ILSs developed by pre-service teachers, which in turn have practical implications for training pre-service teachers who will be designing computer-supported inquiry learning activities.

Keywords Simulation · Inquiry Learning Spaces (ILSs) · Pedagogical design · Pre-service teachers

M. Papaevripidou (✉) · N. Xenofontos · T. Hovardas · Z. Zacharia
Research in Science and Technology Education Group, Department of Educational Sciences, University of Cyprus, 75 Kallipoleos St, 1678 Nicosia, Cyprus
e-mail: mpapa@ucy.ac.cy

Y. Cai et al. (eds.), *Virtual and Augmented Reality, Simulation and Serious Games for Education*, Gaming Media and Social Effects,
https://doi.org/10.1007/978-981-16-1361-6_8

1 Introduction

Pedagogical design when developing learning activity sequences has been emphasized as a topic of primary importance in pre-service teacher training (e.g., [3, 4]). The present study concentrated on learning activity sequences organized around STEM simulations. Specifically, we used resources available in the Go-Lab ecosystem (https://www.golabz.eu/), which include virtual laboratories offering multiple simulation capabilities (https://www.golabz.eu/labs) and applications in the form of software scaffolds (https://www.golabz.eu/apps), in order to train pre-service teachers to design learning activity sequences. The latter had the form of inquiry learning spaces (ILSs, https://www.golabz.eu/spaces) and were authored using the Graasp authoring tool (https://graasp.eu/). The Go-Lab ecosystem and the Graasp authoring tool allow educators to select a virtual laboratory and start designing an ILS around it, by embedding software scaffolds to support student inquiry (see [1]). The perspective taken in this design template stems from good practice in designing inquiry-based learning environments (e.g., [8]), as properly adapted for computer-assisted learning (e.g.,[6, 10]).

Although we have previously evaluated several resources available in the Go-Lab ecosystem (e.g., [2, 5, 9], the present study focused on a more holistic approach from a design perspective. Our main objectives were to examine strengths and weaknesses of pre-service teachers' pedagogical designs, as reflected in the ILSs they delivered during course assignments and in the final exams for two undergraduate university courses. To do so, we introduced pre-service teachers at the University of Cyprus, Department of Education, to the Go-Lab ecosystem and the Graasp authoring platform. After inspecting and elaborating on a pilot ILS, pre-service teachers were requested to design their own ILS in a context chosen by them from the subject domains in the national curriculum for primary school science education. The second set of ILSs was developed by pre-service teachers during the final exams for the courses, where they had to transform a lesson plan (an activity sequence in the form of a learning activity sheet) into an ILS. Using a list of pedagogical design criteria, we assessed the ILSs developed during the course assignment and those designed by pre-service teachers during their exam and compared the scores to outline which design aspects pre-service teachers retained in the inter-contextual design challenge represented by the exam design task (the context of the course assignment was different from the exam context). Specifically, we concentrated on good design aspects that may have matched across the two ILSs delivered by each pre-service teacher. In the final section of this contribution, we discuss the implications of our findings for teacher training in designing computer-supported inquiry learning activities.

2 Methods

2.1 Participants

The study involved pre-service teachers in two courses at the University of Cyprus in the Department of Education. The first course was available for students in the fifth semester of their studies (The Teaching of Natural Sciences; n = 19), while the second course was available in the seventh semester (Computer Science Applications in the Teaching of Science in Elementary School; n = 16). A substantial majority in the sample were females (28 out of 35 students, 80.0%), which is attributed to the well-established preference of female students at the end of secondary education in Cyprus for educational studies. Participation in the study was anonymous and voluntary, and students were given the option to withdraw at any stage of data collection. Data handling and storage were limited to the authors.

2.2 Procedure

Teacher training included presentation of the Go-Lab ecosystem (https://www.golabz.eu/), demonstration of the Graasp authoring tool (https://graasp.eu/), and working with a pilot inquiry learning space. As part of the requirements for both courses, pre-service teachers were asked to develop their own ILS in a context aligned with the national curriculum for primary school science education, and as part of the final exam for the course, they were also asked to transform a lesson plan (an activity sequence in the form of a typical learning activity sheet) into an ILS. The exam design task presented an inter-contextual challenge, where pre-service teachers were asked to design and deliver an ILS set in a different context from the one encountered in their course assignment. The training provided by the course instructors (first and second authors), the course assignment, and the exam design task was identical across the two courses.

2.3 Data Sources and Data Analysis

ILSs developed by the students in response to their course assignment (course ILSs) and ILSs developed by the students during their exam (exam ILSs) were scored according to a list of pedagogical design criteria (Table 1). These had been adapted by Leemkuil et al. [7] and were reformulated and enriched by the authors to address the objectives of the current study. The list with design criteria was not explicitly taught during the courses, but all relevant aspects were incorporated in teacher training. We expected that design skills reflected in the course ILSs would be—at least partially—transferred to the exam ILSs.

Table 1 Criteria for designing inquiry learning spaces (ILSs), with median scores for ILSs delivered by pre-service teachers during course assignment and exam

Category	Criterion	Median for course assignment	Median for exam
Use of simulation	The simulation chosen can support the learning activity sequence in the ILS	2	2
	Students get familiar with the simulation before using it for experimentation	1	0
	The content of the ILS is related to the variables that can be manipulated in the simulation	1	1
Prior knowledge	Student prior knowledge is activated	2	1
	Key necessary prior concepts are explained	1	0
	The content of the apps is adjusted to account for student prior knowledge	1	1
Hypothesis scratchpad (HS)	The HS is placed properly in the learning activity sequence	2	2
	The HS contains variables that can be manipulated in the simulation	2	2
	The HS and its instructions are configured in alignment with the EDT and its instructions	2	1
Experiment design tool (EDT)	The EDT is placed properly in the learning activity sequence	2	1
	The EDT contains variables that can be manipulated in the simulation	2	2
	The EDT and its instructions are configured in alignment with the simulation used	2	1
Scaffolds	Students are given clear instructions/worked examples for how to undertake tasks	1	1
	Hints are used to provide additional support when needed	2	1
	Irrelevant visual or auditory information is not used	2	1
Sequencing	Navigational support/hints are offered to students on how to proceed after completing a task	2	1
	Elements that belong together conceptually are placed together in the sequence	1	1
	The ILS remains focused on the main elements (e.g., key concepts/variables under study)	1	1
Conclusion tool	The conclusion tool is properly placed in the learning activity sequence	2	2
	Students are guided to retrieve their previous work to confirm or reject their hypotheses in the conclusion tool	1	1
	Students are guided to reflect on their previous work	2	0

Scores for criteria were 0 = not addressed; 1 = partially addressed; 2 = fully addressed

The complete list includes 21 criteria, in total, which are grouped into different categories to reflect different design demands, namely the effective use of a simulation within the ILS, the proper consideration of student prior knowledge in the ILS design, the effective placement and configuration of main applications, such as the hypothesis scratchpad (https://www.golabz.eu/app/hypothesis-scratchpad), the experiment design tool (https://www.golabz.eu/app/experiment-design-tool), and the conclusion tool (https://www.golabz.eu/app/conclusion-tool), the inclusion of further scaffolds for students, such as instructions and hints, and the proper sequencing of certain elements in the ILS.

For each criterion, an ILS could receive a score of 2, if its design fully followed good practice as indicated by the criterion; a score of 1, if it partly reflected good practice; and a score of 0, if it did not reflect good practice at all. Two independent coders scored a random selection of 10% of the ILSs using the pedagogical design criteria in Table 1. Inter-rater reliability was slightly over 85% agreement, while mismatches between the coders were settled through discussion.

We computed median values for each individual criterion and means for aggregate scores for the categories of criteria, for course and exam ILSs. We performed Wilcoxon signed ranks tests to investigate significant differences between scores in course and exam ILSs, for each category of criteria. We calculated Spearman's rho correlations between aggregate scores for the categories of criteria in course and exam ILS. A positive correlation would indicate that good practice in pedagogical design was retained by pre-service teachers in the transition between training (course ILS) and exams (exam ILS). Finally, we calculated the difference between total scores for course and exam ILSs, and then, we conducted a tree analysis to outline the individual criteria that accounted for that difference. This analysis would identify the design aspects with crucial importance for the overall performance of pre-service teachers in retaining trained pedagogical design skills.

3 Results

3.1 Differences in Mean Aggregate Scores for Categories of Criteria Between Course and Exam ILSs

The exam ILSs received lower mean aggregate scores as compared to course ILSs for three categories of criteria: use of simulation, prior knowledge, and conclusion tool (Table 2). This indicates that these design aspects were the most difficult to retain in the inter-contextual transfer of design skills.

A closer inspection of median scores for individual criteria revealed specific aspects that need to be highlighted in future pre-service teacher training (see Table 1). For instance, the use of simulations in an ILS should have been accompanied by a familiarization process for students, who need to become familiar with the core features of the simulation before using it for experimentation later on

Table 2 Differences in mean aggregate scores for categories of pedagogical design criteria between the course assignment and the exam (n = 35)

Category of criteria	Mean aggregate score for course ILSs	Mean aggregate score for exam ILSs	Wilcoxon signed ranks test (Z)
Use of simulation	3.94	3.20	-2.29^{*}
Prior knowledge	3.86	2.29	-3.94^{***}
Hypothesis scratchpad (HS)	4.51	4.30	−0.79
Experiment design tool (EDT)	3.63	4.06	−1.44
Scaffolds	3.97	4.03	−0.25
Sequencing	4.23	3.86	−1.74
Conclusion tool	3.69	2.46	-3.68^{***}

Note see Table 1 for a complete list of criteria used, the maximum possible aggregate score per category is 6,$^{*}p < 0.05$; $^{**}p < 0.01$; $^{***}p < 0.001$

in their learning trajectory (Table 1, category = use of simulation, criterion = students get familiar with the simulation before using it). For this criterion, the median score for course ILSs was 1, and it dropped to 0 for exam ILSs. For the category of prior knowledge, the scores for the ILS designs by pre-service teachers on the exam were lower than those for the course ILSs for activating student prior knowledge (median score = 2 for course ILSs and 1 for exam ILSs) as well as for explaining key necessary prior concepts (median score = 1 for course ILSs and 0 for exam ILSs). In the category related to the conclusion tool, the major weakness in exam ILSs referred to guiding students to reflect on their previous work in the ILS (median score = 2 for course ILSs and 0 for exam ILSs). For the rest of the categories, the differences were not significant. In some cases, scores for exam ILSs were higher than those for course ILSs, which implies that the former showed better design for these aspects than the latter (see Table 2 for the categories of experiment design tools and scaffolds).

3.2 Correlations Between Aggregate Scores for Categories of Criteria for Course and Exam ILSs

The total aggregate score for course and exam ILSs correlated positively (Spearman's rho = 0.69, $p < 0.001$), which indicates that good practice in ILS design, when adopted and elaborated upon during the courses, was likely to be implemented by pre-service teachers during their exams as well, in an inter-contextual fashion. A more focused correlational analysis showed that the above-mentioned effect mainly referred to the various applications used by pre-service teachers in their ILSs, namely the hypothesis scratchpad (Spearman's rho = 0.53, $p < 0.01$), the experiment design tool (Spearman's rho = 0.67,

$p < 0.001$), and the conclusion tool (Spearman's rho = 0.67, $p < 0.001$). Pedagogical design criteria within the category of sequencing elements in the ILS also revealed this facilitating association between course assignment and exam scores (Spearman's rho = 0.48, $p < 0.01$). However, we did not find such an effect for the use of simulations, the consideration of student prior knowledge, and the use of scaffolds (other than applications, namely instructions and hints).

3.3 *Significant Determinants of the Difference in Total Aggregate Score Between Course and Exam ILSs*

Figure 1 presents the pedagogical design criteria that were included in the tree model to account for the difference in total aggregate score between course and exam ILSs.

Overall, this difference was negative (see Node 0; mean = −3.63), which indicates that exam ILSs received a lower aggregate score than course ILSs. Despite this overall trend, a number of students succeeded in having a positive difference (i.e., their aggregate score for the exam ILS was higher than their aggregate score for the course ILS); all of these students had adjusted the content of the applications in their ILSs to account for student prior knowledge (Fig. 1, first split, PKNOW_APP = The content of the apps is adjusted to account for student prior knowledge, right branch, mean = 7.38; $n = 8$). In the next split, the tree distinguished pre-service teachers who took into account in their ILS designs the fact that students needed to get familiar with the simulation before using it for experimentation (Fig. 1, second split, SIM_FAM = Students get familiar with the simulation before using it for experimentation, right branch, mean = −2.11, $n = 9$); these students presented a smaller score difference between course and exam ILSs compared to students who did not take this criterion into account (Fig. 1, second split, left branch, mean = −9.28). The last split of the tree is marked by another criterion linked to student prior knowledge, namely explaining key necessary prior concepts (Fig. 1, third split, PKNOW_CON = Key necessary prior concepts are explained). Again, pre-service teachers who took into account this criterion presented a smaller score difference between course and exam ILSs (Fig. 1, third split, right branch, mean = −7.20, $n = 10$) compared to students who did not take it into account (Fig. 1, third split, left branch, mean = −11.88, $n = 8$).

4 Discussion

The present study showed that good practice in pedagogical design may be retained, at least up to a point, in the transition from pre-service teacher training to an inter-contextual design challenge, which was operationalized in our study by the

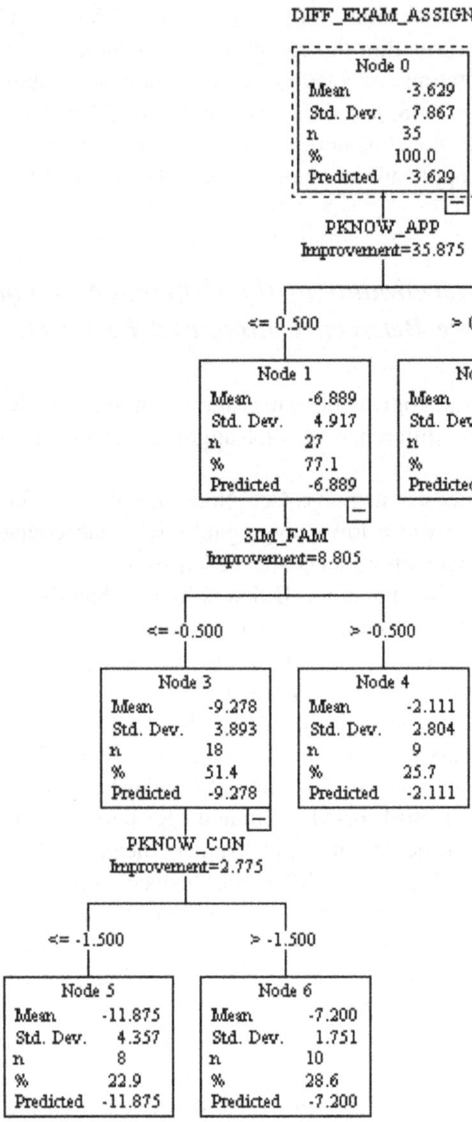

Fig. 1 Tree model for the difference between total aggregate score for course and exam ILSs across the list of pedagogical criteria (Node 0; DIFF_EXAM_ASSIGN; $n = 35$). The mean for the entire sample is negative (−3.63), which indicates that the total aggregate score for exam ILSs was lower than the total aggregate score for course ILSs. Variables depicting pedagogical design criteria (independent variables) are shown at each split together with thresholds for partitioning the sample at each branch (i.e., left and right branches). The tree includes only the criteria that proved significant in the overall partitioning of the sample (see also Table 1): PKNOW_APP = The content of the apps is adjusted to account for student prior knowledge, SIM_FAM = Students get familiar with the simulation before using it for experimentation, PKNOW_CON = Key necessary prior concepts are explained. Each node shows the mean value and standard deviation of the dependent variable (difference between total aggregate score across the list of pedagogical criteria between course and exam ILSs), number of students (n), and percentage of the student sample. Total variance explained by the tree = 79.54%

ILS design task on the exam. Despite the constraints and limitations that could have been imposed on pre-service teachers by the exam setting, they seemed to be quite effective in implementing important pedagogical design competencies, especially in terms of embedding applications such as the hypothesis scratchpad and the experiment design tool in the learning activity sequences they converted to ILSs. Following the same trend, further scaffolds within ILS designs (e.g., instructions and hints) as well as the sequencing of several elements included in their ILSs did not vary significantly compared to ILSs developed during the course assignment. Apart from these strengths in pedagogical design, our findings also revealed a number of weaknesses, which mainly referred to the use of simulations in the exam ILS, the consideration of prior knowledge, and guiding students to reflect on their previous work while using the conclusion tool.

This distinction between strengths (placement and configurations of main applications such as the hypothesis scratchpad and the experiment design tool), on the one hand, and weaknesses in pedagogical design, on the other (allowing students to get familiar with the simulation used for the ILS before using it for experimentation, consideration of student prior knowledge, stimulating student reflection in the conclusion tool), may reflect a distinction between the linear fashion by which inquiry-based learning is usually conceived of (i.e., one-way forward elaboration of learning tasks from an entry point toward an exit point in student inquiry), and the need to integrate in pedagogical design a planned regression to previous stages of student inquiry or a planned process of familiarization with simulations to be used in upcoming stages (see, in this regard, the arrows presented by Pedaste et al. [8] in Fig. 3 on page 56). Pedagogical design in the exam ILSs seems to have been predominantly guided by the main applications available in the Go-Lab ecosystem, which reconstruct the typical sequence of learning activities in inquiry-based learning: the hypothesis scratchpad and the experiment design tool. In contrast, weaknesses were revealed whenever a deviation from this linear onward scheduling of learning routes was needed. These deviations, which are facilitated by multiple developments in computer-supported learning environments, should be a focus of future research and teacher training.

Acknowledgements We thank all our colleagues within the Next-Lab research project (Next Generation Stakeholders and Next Level Ecosystem for Collaborative Science Education with Online Labs), which was funded by the European Community under the Horizon 2020 Call (H2020-ICT-2016-2017; Information and Communication Technologies Call; Grant Agreement: 731685). Our work in Next-Lab inspired and guided this study.

References

1. De Jong, T., Sotiriou, S., Gillet, D.: Innovations in STEM education: the go-lab federation of online labs. Smart Learn. Environ. **1**, 1–16 (2014). https://doi.org/10.1186/s40561-014-0003-6
2. Efstathiou, C., Hovardas, T., Xenofontos, N., Zacharia, Z., de Jong, T., Anjewierden, A., van Riesen, S.A.N.: Providing guidance in virtual lab experimentation: the case of an experiment design tool. Educ. Tech. Res. Dev. **66**, 767–791 (2018). https://doi.org/10.1007/s11423-018-9576-z
3. Hovardas, T.: A learning progression should address regression: insights from developing non-linear reasoning in ecology. J. Res. Sci. Teach. **53**, 1447–1470 (2016). https://doi.org/10.1002/tea.21330
4. Hovardas, T., Zacharia, Z.: A model-based inquiry approach to explore system dynamics and modeling aspects of the prey-predator system. In: Cai, Y., van Joolingen, W., Walker, Z. (eds.) Virtual Reality, Simulations and Serious Games for Education, pp. 99–109. Springer Nature, Singapore (2019)
5. Hovardas, T., Xenofontos, N., Zacharia, Z.: Using virtual labs in an inquiry context: the effect of a hypothesis formulation tool and an experiment design tool on students' learning. In: Levin, I., Tsybulsky, D. (eds.) Optimizing Stem Education with Advanced Icts and Simulations, pp. 58–83. IGI Global, Hershey PA, USA (2017)
6. Hovardas, T., Pedaste, M., Zacharia, Z., de Jong, T.: Model-based science inquiry in computer-supported learning environments: the case of Go-Lab. In: Azad, A.K.M., Auer, M., Edwards, A., de Jong, T. (eds.) Cyber-Physical Laboratories in Engineering and Science Education, pp. 241–268. Springer, Dordrecht (2018)
7. Leemkuil, H., Tsourlidaki, E., Kypriotis, E., Koslowsky, J., Xenofontos, N., Papaevripidou, M., Hovardas, T., Irakleous, M., Zacharia, Z., Dziabenko, O., Siiman, L., Rannastu, M., Mäeots, M., Leijen, Ä., Pedaste, M., Almeida, M.L., Doran, R., Vidal, G., Korventausta, M., van der Graaf, J., de Jong, T.: Releases of learning spaces (40) for primary and secondary (20) education. Next-Lab Deliverable **2**, 6 (2018)
8. Pedaste, M., Mäeots, M., Siiman, L.A., de Jong, T., van Riesen, S.A., Kamp, E.T., Manoli, C. C., Zacharia, Z.C., Tsourlidaki, E.: Phases of inquiry-based learning: definitions and the inquiry cycle. Educ. Res. Rev. **14**, 47–61 (2015). https://doi.org/10.1016/j.edurev.2015.02.003
9. Xenofontos, N., Hovardas, T., Zacharia, Z.C., De Jong, T.: Inquiry-based learning and retrospective action: problematizing student work in a computer-supported learning environment. J. Comput. Assist. Learn. **36**, 12–28 (2020). https://doi.org/10.1111/jcal.12384
10. Zacharia, Z.C., Manoli, C., Xenofontos, N., De Jong, T., Pedaste, M., van Riesen, S.A., et al.: Identifying potential types of guidance for supporting student inquiry when using virtual and remote labs in science: A literature review. Educ. Tech. Res. Dev. **63**, 257–302 (2015). https://doi.org/10.1007/s11423-015-9370-0

Guiding Student Thinking Through Teacher Questioning When Learning with Dynamic Representations

Antti Lehtinen, Markus Hähkiöniemi, and Pasi Nieminen

Abstract Dynamic representations (e.g., dynamic geometry software GeoGebra for mathematics learning and PhET simulations for science learning) offer excellent opportunities for students to conduct investigations and to formulate explanations for the visualized phenomena. In order for this to be effective, students need guidance, for example, for planning their investigations and reflecting on their actions. One way to support students is by prompting them by using questions that are adapted to the students' current situation. This chapter focuses on how pre-service teachers provide guidance for students through questioning and by both structuring and problematizing student learning. Data comes from science lessons taught by pre-service primary school teachers and mathematics lessons taught by pre-service subject teachers. The analysis focused on the different question types the pre-service teachers used as well as how their questioning was adapted to students' situation. The results show how the pre-service teachers used questions both to structure student thinking and to problematize their answers and reasoning. Questioning was not always adapted to the students' needs. We propose that adapting teacher questioning to student thinking requires balancing between structuring and problematizing and high level of interpretation from the teacher. Teachers' skills for interpretation are still beyond the skills of software. Implications for teaching with dynamic representations are discussed.

Keywords Pre-service teachers · Teacher guidance · Inquiry-based learning · Simulations · STEM education

A. Lehtinen (✉) · M. Hähkiöniemi · P. Nieminen
Department of Teacher Education, University of Jyväskylä, Jyväskylä, Finland
e-mail: antti.t.lehtinen@jyu.fi

A. Lehtinen
Department of Physics, University of Jyväskylä, Jyväskylä, Finland

Y. Cai et al. (eds.), *Virtual and Augmented Reality, Simulation and Serious Games for Education*, Gaming Media and Social Effects,
https://doi.org/10.1007/978-981-16-1361-6_9

1 Background and Literature Review

Simulations for science learning and dynamic geometry software for mathematics learning have opened up new possibilities for teaching and learning [2, 4]. As both of them offer a chance for students to interact with the representations and manipulate the different objects and their properties, these are often referred as dynamic representations [1] (opposed to static representations that cannot be manipulated).

Dynamic representations enable the students to experiment with variables and objects and thus discover different scientific and mathematical principles through an active learning process [6]. Unguided inquiry-based learning is criticized for being less effective and efficient than direct instruction for novice learners [11]. The processes involved in inquiry-based learning such as interpreting data and drawing conclusions are difficult for learners without high level of prior knowledge or experience [5]. The students also tend to focus on completing tasks instead of deepening their knowledge [12]. It is clear that guidance (i.e., assistance that aims to simplify, provide a view on, elicit, supplant, or prescribe the reasoning skills involved [13]) must be provided either through the software itself, learning material or by the teacher [15]. Reiser [17] proposes two mechanisms for supporting learning: *structuring* and *problematizing*. Structuring refers to reducing complexity and choice by providing additional structure to the task. This can happen through, e.g., dividing complex tasks into smaller parts or focusing student effort into the parts of the task that are most productive for learning. Problematizing refers to, e.g., eliciting explicit student reasoning by pointing out important distinctions or distinguishing surface-level thinking and disagreements within a group.

One factor that promotes guidance provided by a teacher instead of software or learning material is the teachers' superior possibilities to adapt their guidance to the students' needs [6, 15]. The process of adapting guidance requires the teacher to notice and recognize student thinking [20] and use this information coming from multiple sources including their talk, actions, and even from their body language [18] to guide the students on-the-fly. Then this information needs to be interpreted and used to guide the students' actions and to help them achieve the learning goals. At the same time, the teacher should balance between structuring student work and problematizing it based on their needs [17].

Especially teacher questions can be used both to elicit and probe students' ideas and stimulate productive and higher-order thinking [3] and to provide guidance for the students [19]. Sahin and Kulm [19] distinguish between these two uses for questioning by describing different question types. *Probing questions* are questions that ask the students to explain or elaborate their thinking, use prior knowledge, and apply it to a current problem or idea or ask students to justify or prove their ideas. On the other hand, *guiding questions* ask for a specific answer, ask students to think about or recall a general heuristic or contain a sequence of factual questions to provide ideas or hints that guide toward understanding a concept or completing a procedure. As a third question type, Sahin and Kulm [19] distinguish factual questions that ask for a specific fact or definition, an answer to an exercise or the next step in a procedure.

Lehtinen and Hähkiöniemi [14] studied how pre-service teachers used spaces that a simulation created for explanation to probe for students' explanations through probing questions. Even though the explanations were prompted, they were not always used to adapt the following guidance. Hähkiöniemi [8] found that pre-service mathematics teachers who asked series of guiding questions directed students toward finding an answer through a specific route whereas those who asked series of probing questions elicited learners' thinking and directed them toward forming explanations. These two ways of using questions have a connection with the two mechanisms of supporting learning proposed by Reiser [17]. Adapting guidance can be seen in one part as choosing between problematizing and structuring student learning.

The aims of this chapter are to analyze (1) the processes of structuring and problematizing student learning with dynamic representations and (2) how guidance provided by teacher questioning is adapted to the needs of the students. We examine these instructional practices via two cases from two different contexts where pre-service teachers (PSTs) guide a small group of students—one from upper secondary school mathematics and the other from primary school science. The research question is "How do pre-service teachers structure and problematize student learning with dynamic representations through adaptive questioning?".

2 Methods

We examined data from two sources in order to have richer set of data regarding different ages and different dynamic representations. The first data source is a project about GeoGebra-enriched (www.geogebra.org) inquiry-based mathematics teaching in Finnish lower and upper secondary schools. More details about this project is given in Hähkiöniemi [9]. The second data source is a project about using PhET simulations (https://phet.colorado.edu/en/simulations/) as a part of inquiry-based science teaching in Finnish primary schools. More details about this project is given in Lehtinen and Viiri [15].

2.1 Data Collection

Lessons in both projects were video-recorded. 29 lessons were videotaped for the mathematics education project and 8 for the science education project. One video camera recorded the teachers' actions and talk. The students' actions with the dynamic representation software and the discussions they were having with the PST were recorded either with a hand-held video camera (mathematics lessons) or through a screen capture program (science lessons).

The data analyzed in this chapter are discussions from two of the lessons (one from mathematics and one from science) where a PST guided a small group of students in their investigations. Both lessons contained an introduction, group work, and whole class discussion.

The mathematics lesson chosen for analysis (episodes 1 and 3) was a 90-min 10th grade lesson (short syllabus) about the contingence angle of two tangents to a circle. During group work, students worked in groups of two to four using a computer. They used a beforehand prepared applet Fig. 1 to solve the following tasks:

- What is the size of the angle between the line and the segment CA? What is the size of the angle between the other line and the segment DA?
- How big can the central angle α be, if the location of the point B is changed?
- When are the angles α and β equal?
- Is there some kind of connection between the angles α and β?

The purpose of the task was to engage students in making observations, noticing patterns, and generating conjectures through experimenting with dynamic geometry software and then explaining or justifying these through deductive reasoning. Thus, the task included aspects that are considered as important affordances of using dynamic geometry software in mathematics learning (e.g., [7]).

The science lesson chosen for analysis (episode 2) was a 3rd grade 45-min lesson about the variables that affect the balance of a seesaw. During the group work, students worked in groups of three to five using a computer with a PST guiding them. The students used the "Balancing Act" PhET simulation to first explore balance through exploration with different weights and a seesaw. After this, the simulation provided the students with the following types of tasks:

- Where to place an object with a known mass to the seesaw (where another object with a known mass already is) in order to balance it?
- Which way will the seesaw turn? (Fig. 2)
- What is the mass of an object? (The students had to use an object with known mass and the seesaw to deduce the mass).

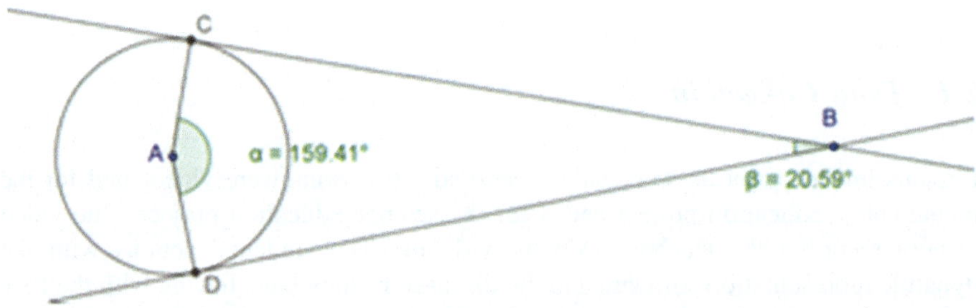

Fig. 1 GeoGebra applet in which one can drag points A and B (see https://ggbm.at/acMDDeae)

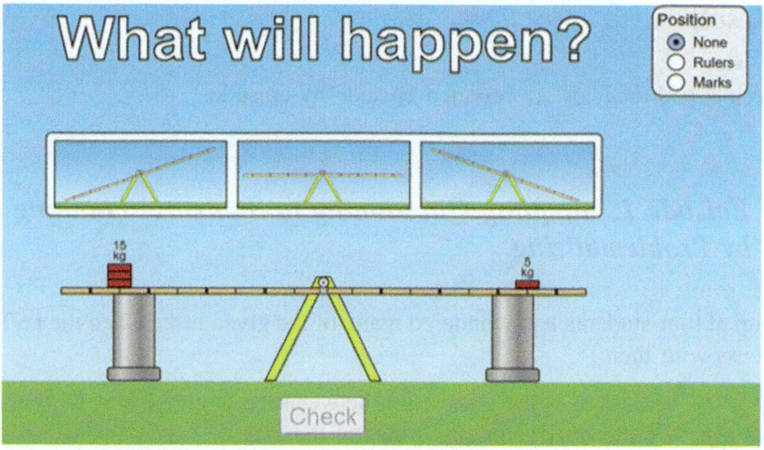

Fig. 2 An example task from the "Balancing Act"–simulation

In principle, the tasks required the students to apply the moment arm rule that they could discover beforehand through exploration. As the students were 3rd graders, they were not able to formulate the rule in its entirety. Instead, they came up with qualitative or simpler formulas that still allowed them to solve some of the tasks.

2.2 *Data Analysis*

From the lesson transcripts, teacher questions were coded based on the categories by Sahin and Kulm [19] : probing, guiding, and factual questions (see definitions in the previous section). Teacher utterances were considered questions if they invited the learners to produce an oral response. For the mathematics lessons the inter-rater agreement was 89% (sample of 150 questions) and Cohen's $\kappa = 0.845$ and for science lessons the inter-rater agreement (for a sample of 101 questions, i.e., 12% of the questions) was 90% and Cohen's $\kappa = 0.832$. After coding the teachers' questions, the transcripts were divided into episodes that were marked by a change in topic, contrast in behavior or transition to the next type of conversation or activity [10]. The episodes were then analyzed based on the features of structuring or problematizing student learning. In the end, three episodes (that came from two lessons) were chosen for further analysis for showcasing three different ways of structuring and problematizing student learning using adaptive teacher questioning.

3 Results

The results and analysis are reported episode by episode.

3.1 *Episode 1: Adapting Questioning to Students' Thinking by Problematizing*

A group of four students had wondered many of the given tasks when the PST came to discuss with them.

1. PST 1: Then, what is the connection between the angles α and β? (Factual)
2. Student: We did not even get the idea of the task.
3. PST 1: Oh well, it means that if they like depend on each other. Is it like…
4. Student A: They go like in the same proportion.
5. PST 1: Yeah, so if you know, for example, α, can you calculate β? (Guiding)
6. Student: Yeah. [Student mumbles something about a square and points to the screen.]
7. PST 1: Yeah, if it is a square, you can calculate that? Isn't it like that? Well, what if it is not a square? Can you still calculate it somehow? (Guiding)
8. Student: You can.
9. PST 1: How? (Guiding)
   [Silence, 15 s. Students drag points and look at the screen.]
10. Student A: Yes you can. Yeah, now I got it.
11. PST 1: So.
12. Student A: Because these. Because these are 90 [points to C and D] and this is 360 in total [points to the quadrangle ADBC], then only calculate these and you will get it from that.
13. PST 1: Uh-hum. Right.
14. Student: Yay.
15. PST 1: But now, there is the question that how did you reason that this is 90°. (Probing)
16. Student A: Well, because this is a circ-. Well okay.
17. Students: [inaudible]
18. PST 1: How did you reason that it is 90°? (Probing)
19. Student: [Mumbles something about a quadrangle.]
20. Student: A quadrangle.
21. PST 1: It is not quite enough for it.
22. Student A: Yeah, it isn't.

In this episode, the PST noticed that the students do not understand the task through questioning (turn 1). This information was used to guide the students through structuring the task into a more closed one (turns 3 and 5). After the students had understood the task, the answer they provided is just for a special situation. The PST problematized this (turns 7 and 9) and the students were then able to provide a more general solution after manipulating the dynamic representation (turn 12). Building on their answer the PST probed the students for their reasoning (turns 15 and 18). When they were unable to provide this, the PST moved on. The question about the justification of the angle between the tangent and radius was a challenging question as several groups had difficulties in the justification. The PST turned to this issue during the closing whole class discussion.

3.2 *Episode 2: Not Adapting Questioning to Student Thinking by Structuring*

The students were completing the tasks that the simulation provided for them. Previously, the tasks had asked them to place a certain weight to the seesaw in order to balance it. The current task asked whether the seesaw would tilt to the left, stay in balance or tilt to the right when there is a 5 kg weight on the left and a 15 kg weight of the right. The distance from the fulcrum was the same for both of the weights.

23. Student B: So now…
24. Student C: We'll have to move this one that way
25. Student B: Yeah we'll have to move this that way so
26. PST 2: That is right but in this one you don't have to move them but if the situation is like this that there is fifteen kilos in that place and this one is here so which one of these will happen (probing)
27. Student C: So in here happens like this that it first goes like this because this one like goes away and then in here it is in balance and in here it kind of goes away
28. PST 2: Mmm but if
29. Student A: I think it is that one. [Student A points to the correct answer]
30. PST 2: Which of these is more—which one is heavier? (guiding)
31. Student C: This one. [points to the 15-kg weight]
32. PST 2: This one—are these on the same line? (guiding)
33. Students B ja C: Yes.
34. PST 2: Yes, so if this one is heavier, then what will happen? (guiding)
35. Student C: It goes down.
36. Student B: It goes there, so that picture.
37. PST 2: Ok, try it and press "Check answer".
38. Student A: Yes, it was.

Again in this episode, at first the students did not understand the task (turns 24 and 25) and tried to manipulate the dynamic representation in a way that was not possible. Compared to the first episode, this time the PST simply verbally prompted the students (turn 26) when they had not understood the task in writing. After the students had understood the task, they were able to provide the right answer but without explaining why it would be the correct one (turn 29). The PST did not problematize this or prompt the students for reasoning for their answer. Instead, he/she structured the students' learning by implicitly providing them with a rule that they could use to solve similar tasks (turns 30, 32, and 34). Through a series of answers, the students were implicitly reasoning for their answer but this was not made explicit for the students, as the question chain did not start from the students' answer.

3.3 Episode 3: Adapting Questioning to Student Thinking by Both Problematizing and Structuring

A student had been mainly working alone although he had been assigned a pair. The student had thought about the tasks when the PST started to discuss with him.

39. PST 1: How about then this, how big can the central angle be if you drag *B*? (Factual)
40. Student: Well, it varies greatly. If you drag it here, to the far end [points to the corner of the window], then it would be 163.45. If it would be very close, like here, then it would be 9.33 [points close to the circle] and then here it would of course be 360 [moves a finger along the circle].
41. PST 1: How about, take, uh-hum. Let's move this. Now you can drag it more and more. [Teacher zooms out and drags the point B farther away from the circle.]
42. Student: Oh, you can do like that
43. PST 1: Uh-hum. When would this be the biggest? You can drag this pretty far away. You can place it as far as you like. Nevertheless, what would the central angle bee when it is the biggest? (Guiding)
44. Student: Isn't it 360.
45. PST 1: Uh. This. So.
46. Student: Oh.
47. PST 1: Like the angle between these [shows the angle]. (Guiding)
48. Student: I would say that it is 180 here [makes a gesture of 180°].
49. PST 1: Can it be 180? (Guiding)
50. Student: [Silence, 4 s.] Yes, I think that it can be.
51. PST 1: Uh. So this would be 180. How would these two lines go then [points to the tangent lines]? If some line would intersect them like that? (Guiding)
52. Student: Like parallel?
53. PST 1: Yeah. If they are parallel, then what would happen to this point here? If these two are parallel and this is the intersection point of them. (Guiding)
54. Student: Then it could not be there because these go like in parallel.
55. PST 1: Yeah, right. So it cannot newer be 180, can't it.
56. Student: So it cannot be 180.

57. PST 1: But what, yeah. It will, however, get very close to it, wont it.
58. Student: Yeah.

In this episode, the student's difficulty was initially not about understanding the task but about how to use the dynamic representation to investigate. The teacher noticed this from the student's incomplete answer (turn 40) and showed how to use the software (turns 41 and 43). After this, the student had difficulties in recognizing the correct angle to study (turn 44). The PST showed him/her the correct angle (turn 47). This information and correct manipulation of the representation enabled the student to answer the task (turn 48). The answer was almost correct. The PST noticed and problematized this. The student's answer was confirmed with a guiding question (turn 49). After this, the PST structured student reasoning through a series of guiding questions (turns 51 and 53) building on their original answer.

4 Discussion

The results of this study show how teacher questioning *can be* but *is not always* adapted to student thinking when using dynamic representations. The analyzed episodes differed in both the adaptation of teacher guidance and how the student learning was structured or problematized. In all of the episodes, the students misunderstood the assigned tasks in many ways. In the first episode, the students did not understand what kind of connection they were asked for. In the second episode, the students did not notice that the task differed from the previous one. In the third episode, the student did not understand how he/she was asked to use the dynamic representation to solve the task. To respond to these difficulties, in all of the episodes, the PSTs first interpreted the difficulty and then tailored questions to help the students to understand the task properly. The help was offered in the form of questions which rephrased the task to be more concrete (episode 1), rephrased the task to clarify how the use of dynamic representation differed from a previous task (episode 2) or first showing how to use the dynamic representation appropriately (episode 3). We interpret that in these instances the teachers adapted their questioning to the specific difficulty the students were confronting. This is similar to Hähkiöniemi et al.'s [7] study, in which a teacher either narrowed or widened the starting situation of an open problem depending on the students' work so far. In addition, a crucial feature of adapting teacher questioning to students' difficulties in understanding a problem is to just help them to understand the problem but not unnecessarily reduce the cognitive demand as sometimes happens when challenging problems are implemented [21].

After the students understood the tasks, there were still several issues that the PSTs needed to interpret and react to. In the first episode, students first proposed only a special case for a solution, and finally, when prosing a general solution, their justification needed reworking. In the second episode, the students gave only an answer without underlying reasoning. In the third episode, the student had another flaw in understanding the problem and, after understanding the problem, suggested a slightly incorrect answer. In the first and third episodes, the PST raised the issues that needed more consideration and prompted students to think about them either through probing or guiding questions. On the contrary, in the second episode the PST straight away asked a sequence of guiding questions that implicitly provided the students with a way to reason for their answer. Thus, the PST was using questions to structure [17] students' thinking whereas in the other episodes, the PSTs were using questions to problematize [17] thinking. We propose that *adapting teacher questioning is balancing between structuring and problematizing and requires high level of interpretation.* In the first and third episodes, questioning is adapted as the problematic issues are problematized. However, in the second episode, questioning is not adapted because the PST is structuring students' thinking without basing it on students' own reasoning. The PST does not make an effort to interpret student thinking and whether structuring was needed. Of course, this does not imply that structuring means always that the questions are not adapted. For example, in the third episode, the PST adapted questioning by structuring the student's thinking because after problematizing the student was still adhering to an incorrect idea. The difference there was that the PST made an effort to interpret student thinking via first problematizing by questioning and thus the following structuring was based on the student's knowledge.

We have proposed here that adapting questions to students' current situation in a dynamic representation is an essential feature in guiding students' learning. Moreover, we have proposed that adapting questions happens through interpreting student thinking and then balancing between structuring and problematizing through questioning. This resonates with formative assessment discussions in which teachers probe for information about student learning and then use this information to promote learning [16, 18]. Both concepts emphasize teacher noticing [20] where that teacher is continuously scanning the lesson to recognize important events. In adapting questions, student idea is recognized, then interpreted and an appropriate question is asked. Previous studies have suggested benefits of teachers basing their guidance on high level interpretation of student work [20]. This is something that is difficult to program into the software. For example, software might prompt students to respond *whether* they understand a problem, but a teacher can interpret *how* the students understand a problem. Getting the answers to these "how" questions demand a lot of interpretation capabilities and as discussed in the introduction, teachers have (at the moment) better capabilities for this than software does.

Acknowledgements This work has been funded by the Academy of Finland (project number 318010).

References

1. Ainsworth, S., van Labeke, N.: Multiple forms of dynamic representation. Learn. Instr. **14**(3), 241–255 (2004)
2. Cai Y., van Joolingen W., Walker Z. (eds.). VR, Simulations and Serious Games for Education. Gaming Media and Social Effects. Springer. (2019)
3. Chin, C.: Teacher questioning in science classrooms: approaches that stimulate productive thinking. J. Res. Sci. Teach. **44**(6), 815–843 (2007)
4. de Jong, T.: Technological advances in inquiry learning. Science **312**(5773), 532–533 (2006)
5. de Jong, T., van Joolingen, W.R.: Scientific discovery learning with computer simulations of conceptual domains. Rev. Educ. Res. **68**(2), 179–201 (1998)
6. de Jong, T., Lazonder, A.W.: The guided discovery learning principle in multimedia learning. In: The Cambridge Handbook of Multimedia Learning, pp. 371–390 (2014)
7. Hähkiöniemi, M., Leppäaho, H., Francisco, J.: Teacher-assisted open problem-solving. Nordic Stud. Math. Educ. **18**(2), 47–69 (2013)
8. Hähkiöniemi, M.: Using questioning diagrams to study teacher-student interaction. In: Silfverberg, H., Kärki, T., Hannula, M.S. (eds.) Nordic Research iIn Mathematics Education: Proceedings of NORMA14, Turku, June 3–6, 2014 pp. 91–100. Finnish Research Association for Subject Didactics, Turku, Finland (2015)
9. Hähkiöniemi, M.: Student teachers' types of probing questions in inquiry-based mathematics teaching with and without GeoGebra. Int. J. Math. Educ. Sci. Technol. **48**(7), 973–987 (2017)
10. Jordan, B., Henderson, A.: Interaction analysis: foundations and practice. J. Learn. Sci. **4**(1), 39–103 (1995)
11. Kirschner, P.A., Sweller, J., Clark, R.E.: Why minimal guidance during instruction does not work: an analysis of the failure of constructivist, discovery, problem-based, experiential, and inquiry-based teaching. Educ. Psychol. **41**(2), 75–86 (2006)
12. Krajcik, J., Blumenfeld, P.C., Marx, R.W., Bass, K.M., Fredricks, J., Soloway, E.: Inquiry in project-based science classrooms: initial attempts by middle school students. J. Learn. Sci **7** (3–4), 313–350 (1998)
13. Lazonder, A.W., Harmsen, R.: Meta-analysis of inquiry-based learning: effects of guidance. Rev. Educ. Res. **86**(3), 681–718 (2016)
14. Lehtinen, A., Hähkiöniemi, M.: Complementing the guidance provided by a simulation through teacher questioning. In: Proceedings of the Annual Symposium of the Finnish Mathematics and Science Education Research Association 2015, ISBN 978–952–93–8233–0. Matematiikan ja luonnontieteiden opetuksen tutkimusseura ry (2016)
15. Lehtinen, A., Viiri, J.: Guidance provided by teacher and simulation for inquiry-based learning: a case study. J. Sci. Educ. Technol. **26**(2), 193–206 (2017)
16. Nieminen, P., Hähkiöniemi, M., Viiri, J.: Forms and functions of on-the-fly formative assessment conversations in physics inquiry lessons. Int. J. Sci Educ, 1–23 (2020)
17. Reiser, B.J.: Scaffolding complex learning: the mechanisms of structuring and problematizing student work. J. Learn. Sci. **13**(3), 273–304 (2004)
18. Ruiz-Primo, M.A.: Informal formative assessment: the role of instructional dialogues in assessing students' learning. Stud. Educ. Eval. **37**(1), 15–24 (2011)
19. Sahin, A., Kulm, G.: Sixth grade mathematics teachers' intentions and use of probing, guiding, and factual questions. J. Math. Teacher Educ. **11**(3), 221–241 (2008)
20. Sherin, M., Jacobs, V., Philipp, R. (eds.): Mathematics Teacher Noticing: Seeing Through Teachers' Eyes. Routledge (2011)
21. Stein, M.K., Grover, B.W., Henningsen, M.: Building student capacity for mathematical thinking and reasoning: an analysis of mathematical tasks used in reform classrooms. Am. Educ. Res. J. **33**(2), 455–488 (1996)

Bringing Simulations to the Classroom: Teachers' Perspectives

Koen Veermans and Tomi Jaakkola

Abstract The potential value of simulations in education has been argued already for many years, with the arguments often referring to learning, motivation, or both. However, implementation of educational simulations in classrooms is still lower than what policy makers and researchers would like to see. This study is a continuation of a series of empirical studies where students have investigated the basic principles of electric circuits in a simulation environment in controlled research settings. These studies show positive effects of the simulation both in terms of learning outcomes and interest. The present study focuses on the same simulation environment but changes the focus from students to teachers in regular classrooms. The participants were nine Finnish elementary school teachers. They implemented the circuit simulation in their regular classrooms and reflected on their and their students' experiences regarding the implementation. The overall picture that emerges is that for most teachers this was a positive experience, but the results also reveal useful pointers and areas of improvement to take into consideration when thinking about bringing simulations to the classroom. Both the positive experiences and the useful pointers will be discussed in the broader framework of bringing simulations to the classroom.

Keywords Computer simulation · Inquiry learning · Teacher · Learning environment

K. Veermans (✉)
University of Turku, Turku, Finland
e-mail: koevee@utu.fi

T. Jaakkola
Tampere University, Tampere, Finland
e-mail: tomi.jaakkola@tuni.fi

Y. Cai et al. (eds.), *Virtual and Augmented Reality, Simulation and Serious Games for Education*, Gaming Media and Social Effects,
https://doi.org/10.1007/978-981-16-1361-6_10

1 Introduction

Computer-based simulations or virtual labs—wherein students use interactive simulation software to conduct hands-on experiments by virtually manipulating virtual equipment and materials—have become an increasingly popular alternative to real equipment in science classrooms. Probably, the main reason behind the growing popularity of simulations is their improved usability and availability. Dedicated online repositories such as phet.colorado.edu (PhET) and golabz.eu (Go-Lab) offer hundreds of simulations that can be instantly deployed at school or at home. Modern simulations work directly in an Internet browser; to use them, only a PC or mobile device with an Internet connection is required. In last spring's pandemic situation, simulations—as well as other online resources for learning and teaching—came into exceptional need as schools around the world were forced to switch to distance learning and look for alternatives to traditional instruction.

One of the theoretical background ideas of using simulations and real labs in science education is that learners should have an active role in the learning process [1]. Traditional teacher-led science instruction has been accused of on the one hand not activating, challenging and inspiring pupils sufficiently [2] and on the other hand resulting in inert and disconnected knowledge [3]. There is robust evidence showing that students learn better when they have an active role in the learning process, and learning is connected to their prior knowledge [4]. In this study, we use the concept of inquiry learning to refer to active learning and student-centered pedagogy. A general starting point for inquiry learning is that answers to the tasks or the key laws of the phenomenon under investigation are not directly accessible to learners, instead, learners are encouraged to conduct investigations under teachers' guidance and set up experiments in the subject matter, in order to seek and find out (or discover the answers and solutions to the tasks and problems). While conducting the experiments, learners also gain valuable first-hand experience of activities that mimic aspects of scientific research, and they develop competences in relation to twenty-first century skills that have been put forward in major policy documents (e.g., [5, 6]).

In addition to practical reasons, the use of simulations in science education is supported by the good learning outcomes obtained in research. According to research, the general trend is that learning with simulations results in at least as good (e.g., [7, 8]) and sometimes even better (e.g., [9–11]) learning outcomes than when learning involves real equipment. Learning in a simulation environment would also seem to motivate students [12]. It appears that in those studies where simulations have produced better learning outcomes than real equipment, the special features of the simulations have improved learning outcomes. Finkelstein et al. [10], for example, conducted a study during a university physics course where students studied the basic principles of electric circuits using a simulation or real equipment. Contrary to real circuits, the simulation enabled students to observe continuously the electron flow inside the circuits. The outcome of the study was that the students using the simulation outperformed the students using the real equipment on a conceptual knowledge test administered at the end of the course. According to the authors of the study, the fact that the simulation provided direct perceptual access to the concept of current flow helped students to gain better understanding of the functioning of DC circuits. What is particularly interesting in Finklestein's results is that students that learned with the simulation also succeeded better than the real equipment group in a transfer task where they had to assemble real circuits and explain how those worked. In this context, however, it should be emphasized that the primary learning objective with simulations is a conceptual understanding of the phenomenon under investigation and not, for example, the learning of practical laboratory skills. It is also important to understand that the way how learners make the discoveries in a simulation environment is fundamentally different from how scientists initially discovered these things (e.g., electron flow is not directly observable in any lab) [13]. In previous studies on educational simulations, the main focus has been on students and their learning outcomes, which is well justified as the main aim is to develop instructional methods and learning environments that promote learning and engage students. In these studies, a researcher typically acts as a teacher with the aim of having more control of the impact of individual teacher on the results of the experiment, thus leaving the real teacher outside of the research context. Because it is ultimately the teacher that decides whether or not to use simulations in the classroom and who will be responsible for the implementation, it is also important to consider their views as this may provide insights that may also benefit designers. Therefore, the main focus in the current study was on teachers' perspectives and experiences related to the implementation of a simulation learning environment in their classroom.

2 The Present Study

The present study was carried out in Finland as part of a European project aiming to advance inquiry-based science education in schools across Europe. For this, a training course was developed and organized for teachers to improve their knowledge base in inquiry learning and to provide them with hands-on experiences of inquiry learning. As part of the training course, the teachers' task was to implement three distinct inquiry lessons with their pupils. This paper focuses on teachers' experiences from the first inquiry lesson that was implemented using a simulation learning environment.

A training session was held for teachers about a week before the simulation lesson. Presented training materials were shared with teachers at the end of the session, to allow teachers to study the materials at their own pace to prepare for the upcoming lesson. At the beginning of the training session, teachers were told about the theoretical starting points of inquiry learning and introduced to the Pedaste et al. [14] model of inquiry learning. The model conceptualizes inquiry learning as a cyclical process that consists of five interconnected phases. The model starts with the orientation phase that introduces the topic and aims of the study to stimulate learners' curiosity. In the conceptualization phase, learners state hypotheses and/or research questions, which should guide their actions in the following investigation phase when learners gather empirical evidence and conduct experiments to answer the research questions. In the conclusion phase, findings from the investigation phase are interpreted and summarized. The discussion phase is directly connected to all the other phases and is thus an integral part of the whole inquiry cycle. At the end of the inquiry cycle, discussion refers to communicating the outcomes of the study to outside parties (e.g., peers) and to reflecting on the overall inquiry processes. The inquiry learning model was presented to teachers to help them to see the process of inquiry learning as a continuous process that consists of different phases.

Next, the teachers were introduced to the simulation and related worksheets to be used in the upcoming lesson. It was a circuit simulation (Fig. 1) that allows students to learn about electric circuits by building circuits, conducting electrical measurements (e.g., potential difference across bulbs) and observing circuit behavior (e.g., the brightness of bulbs). The simulation is a HTML5 application that works on any modern Web browser, both on PC and mobile. As can be seen in the figure, the representation level of the simulation is semi-realistic: The circuit layout is schematic, but the brightness of the bulbs changes dynamically based on the circuit configuration. The simulated model is authentic except that the wires have no resistance and the battery is ideal (i.e., it always remains fully charged).

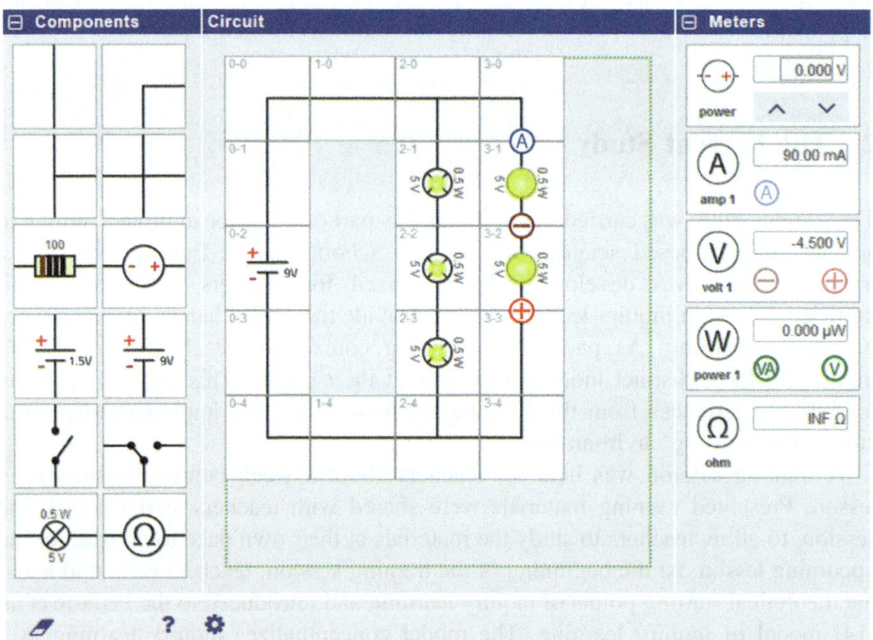

Fig. 1 Circuit simulation (https://www.golabz.eu/lab/electrical-circuit-lab)

The worksheets scaffold students' inquiry process in the simulation learning environment. There are nine worksheets in total that progressively become more difficult (more complex circuits) and open-ended (less support). However, the basic structure of the worksheets is the same, and together they create many small inquiry cycles as can be seen in the worksheet example in Fig. 2. To activate students' prior conceptions, they are asked to predict circuit behavior before actually building the circuit with the simulation. This corresponds to the conceptualization phase in Pedaste et al. [14] model, which is then followed by the investigation phase when students build the circuit and conduct requested measurements. At the end of the worksheets, students are asked to summarize their findings (conclusions phase). The orientation and discussion phases are also part of the activity (but not explicitly part of the worksheets): In the beginning of lesson, teacher introduces the topic and demonstrates the basic functioning of the simulation, and as pupils work in pairs throughout the lesson, this means that the discussion phase is an integral part of each worksheet. The teacher mainly monitors the work of the pairs and helps them when needed.

The above worksheets and corresponding simulation have been tested previously with upper elementary and lower secondary students (e.g., [15–19]. The outcomes of these studies have been largely positive: As can be seen in Fig. 3, the simulation was able to improve students' understanding of electric circuits over a range of five grades (4th–8th; 9–15 years), and students' interest levels were from above average to high, though there seemed no clear relation between the level of interest and learning outcomes and/or learning gains. In the present study, we wanted to move the focus from students to teachers and investigate how teachers perceive the simulation activity after they have implemented it in their classroom.

For this purpose, teachers ($N = 9$) were asked to complete a questionnaire after the simulation activity. The questionnaire consisted of 16 semi-structured questions that were organized along three broad themes: thoughts and expectations prior to the implementation of the simulation activity, experiences during the implementation and the connection of the activity with the inquiry model. The analysis uses a combined deductive and inductive framework [20] where the answers to these questions will first be analyzed deductively following these three themes and their respective questions followed by a discussion that highlights themes that arise inductively from this analysis.

3 Analysis and Results

One of the first things that becomes clear from the answers to the questions is that there is no two teachers or classrooms that are the same. On practically all questions, there was variety in the thoughts and experiences that the teachers shared.

a) WITHOUT COMPUTER Predict what happens to the brightness of the bulbs if a second bulb is added to the previous circuit as illustraded in the nearby pictures (2 bulbs in series). Check if you agree

☐ There is no change in the brightness of the bulbs

☐ The bulbs become brighter

☐ The bulbs become dimmer

☐ The upper bulb is brighter than the lower bulb

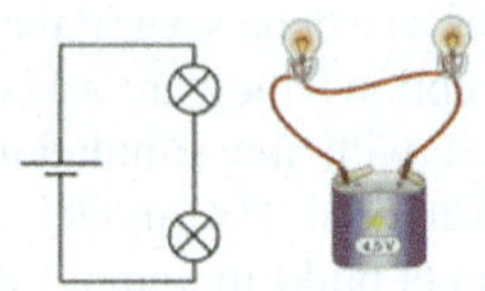

b) WITH COMPUTER Test your prediction by constructing the circuit with the computer. What do you notice (check if you agree)?

☐ There is no change in the brightness of the bulbs

☐ The bulbs become brighter

☐ The bulbs become dimmer

☐ The upper bulb is brighter than the lower bulb

c) WITH COMPUTER Measure the voltage of the upper bulb of the circuit with the voltage meter as instructed in the nearby picture

The voltage of the upper bulb is ________ volts.

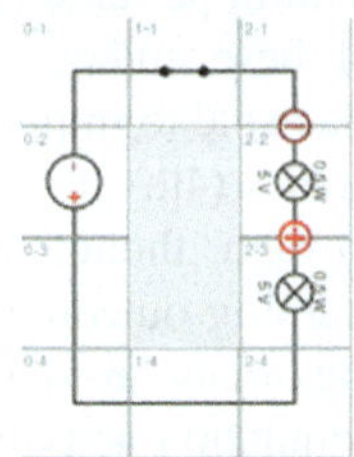

d) WITH COMPUTER Now, measure the voltage of the lower bulb. The voltage of the lower bulb is___volts.

e) What can you say about the voltages of the two bulbs?

☐ The voltages are the same ☐ The voltages are different

What can you conclude about the bulb voltages in comparison to the voltage of the battery in a two-bulb series circuit (check if you agree with a statement)?

☐ The voltage of a single bulb and the voltage of the battery are the same
☐ The voltage of a single bulb is half of the voltage of the battery
☐ The voltage of the battery is divided equally between both bulbs
☐ The total voltage of two bulbs is equal to the voltage of the battery
☐ The total voltage of two bulbs is half of the voltage of the battery
☐ The total voltage of two bulbs is two times the voltage of the battery

Fig. 2 Example worksheet with instructional support

3.1 Before the Implementation

The first questions tapped into the ideas that teachers had before the actual implementation of the simulation lesson. Probably because teachers did not know what questions would follow, some of the answers also already contained references to the actual implementation. The first question asked the teachers how they

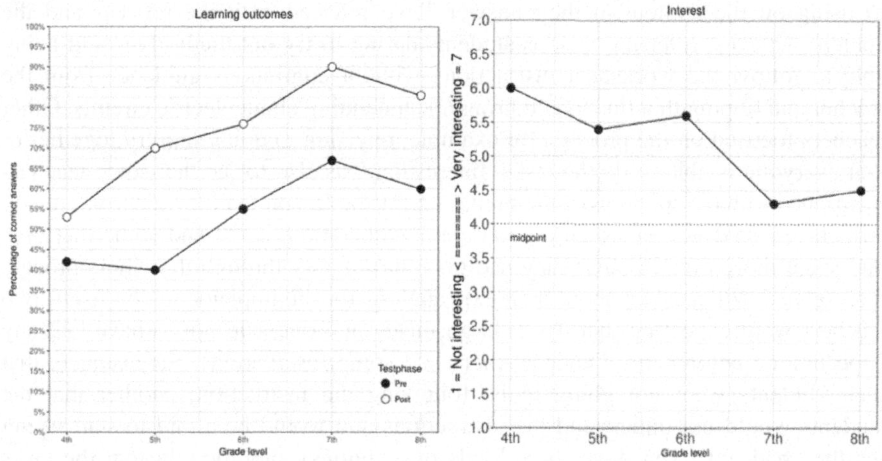

Fig. 3 Trends in learning outcomes (pre- and post-) and interest across grades from the experimental studies

saw teaching electricity circuits (easy or challenging). Most teachers said that they see it as challenging with arguments that refer to themselves (own knowledge is weak), the students (it may feel distant for students) or resources (we do not have proper equipment). Two teachers said it is both easy and challenging, and only one considered the topic easy. The next question asked the teachers about their first impressions of the worksheets that accompanied the simulation. Again, there was a variety in the answers of the teachers. Some teachers were leaning more toward calling the worksheets clear—one teacher calling them clear, one also clear but adding that the worksheets sometimes required a lot of reading (which students do not always do), one interesting, one that they looked difficult but turned out to be well structured—whereas others saw them as challenging or difficult (e.g. one teacher referring to them as challenging and another as messy and difficult). On the question whether they felt the instructions in the worksheets were clear, several teachers said that they thought so themselves, but sometimes adding some disclaimers from what they saw in the classroom during the implementation, like students having difficulty with the measurements, that careful reading was required, that some had quite a bit of text and could have been "cut" up into smaller units, that this need for careful reading and the amount of reading gave difficulties for 2nd language students, but also that they became less self-explanatory toward the end, and that that was a good thing because it made that students needed to "know" things from before. Their impression of the simulation itself was the one of the few things where the teachers generally agreed, with six teachers expressing very positive comments adding notions like easy, motivating, interesting and fun. The next question about the objectives again brought much more diversity and also illustrates the general "dilemma" of inquiry being about the content or the process. Four teachers explicitly referred to learning something about electricity, thus focusing on the content in their answer. Two referred both the process and the content with one referring to it as students daring to try out themselves and being able to follow the worksheet instructions without continuous guidance from the teacher and hoping that this would bring understanding about electric circuits. Other teachers focused on the process, for example, referring to doing inquiry together or that everyone is able to do the tasks. Interesting was also to see that some teachers mentioned to have spent less time preparing for the lesson than for a normal lesson (which one said was not such a good idea), some about equal and some that they had spent more time because they had gone thoroughly through the whole process themselves. When asked about their expectations for the implementation itself, two teachers were expecting that it would go nicely and one said not to have had any expectations beforehand. Other teachers had some reservations, for instance, that their students were not going to be following the instruction, another that the students would not understand the instructions and would be afraid to start trying (as they had not often done these kinds of activities), one thought that the tasks would be hard for the students but they turned out not to be, and one thought that giving students guidance would logistically be a problem but that turned out to be compensated by students working in pairs and helping each other, thus off-loading some of the work from the teacher.

3.2 *During the Implementation*

The next set of questions asked the teacher about experiences during the implementation. The clearest consensus was on the question related to the simulation itself. Only one teacher mentioned some issue but still said the simulation worked well. One teacher had apparently used both computer and tablet and mentioned that it worked even nicer on the tablet because things could be dragged and dropped with the fingers.

When asked whether the worksheets worked as they had anticipated, many teachers referred to the active attitude of the students during the implementation. This was however paired with different remarks: It went well, students were trying out and testing things, in the beginning, there were a lot of questions, I needed to guide more than expected because students were not reading instructions and just started doing things, and the pair work did not result in the depth of discussions that I had hoped for. Regarding the pedagogical virtue of using the simulation in the classroom, four teachers answers were very brief (two excellent, one really good and one ok). The others were more elaborate, and three of these teachers explicitly referred to real equipment, but while one said it made it clearer than working with real equipment, the other two had some doubt about the connection between simulation and real equipment. One was not sure if students would be able to do this also with real equipment but also stressed that the affordances of the simulation triggered enthusiasm in learning, and the other was not sure about transfer to real either, but thought that demonstrating them, the real equipment shortly beforehand would help to make the connection between simulation and reality. One teacher noticed differentiation within the classroom with some really learning and others just progressing through the tasks without acquiring in-depth knowledge. Teachers were also asked what aspects they liked in the material, and if there would be things that could be improved. One teacher answered this question with just "to be able to use it" referring to the intention to use the activity also in future. Several others referred to motivation and the eagerness in students as the best parts. Other good things that were mentioned were the brightness of the bulbs, quick and easy measurements, being able to try things out and clear self-contained worksheets. Related to things that could be improved, one of the teachers mentioned that the worksheets could have benefited from clearer, shorter sentences (a similar remark was made by another teacher in relation to second language and special needs children) and that sometimes the relation between pictures and text was not clear. One of the teachers missed some kind of feedback for students, so they could see if they are on the right track or alternatively monitor the students by checking the answers before students would move on to the next task could also be an option (but was not sure how that could be arranged). The final question in this section addressed potential improvement of the material in terms of supporting teachers. This elicited some general remarks (to know these resources exist, to have/take time for these kinds of activities), and some more specific suggestions for improvement (adding a kind of manual or glossary that explained all simulation elements, an "answer key" for worksheets, proper instruction and manual for teachers, and suggesting teachers to emphasize the importance of following the instructions on the worksheets to their students).

3.3 Connection with the Inquiry Model

The last set of questions asked teachers to reflect on the implementation in relation to the five phases of the Pedaste et al. [14] model that was introduced to them during the training. In answering the question that asked which phase was realized most successfully, most teachers mentioned the investigation phase, often in connection with telling that the students were really engaged during the investigation phase and emphasizing the self-directed nature of the activity. Three teachers also mentioned the conclusion, and one the discussions that students had in the process. These discussions were mentioned by four teachers as something that did not work that well. Reasons mentioned were time constraints and difficulties of students with discussing. Three teachers mentioned the orientation, in one case, this came from the (mis)understanding of how to use material, and in the other, it was a deliberate choice to use as much time in the computer classroom for the investigation as possible. The final question asked if they thought this activity gave a good experience with the inquiry model. Most teachers answered positively to this question, but in some cases also added comments on how this could be even improved with two teachers mentioning that going over the entire cycle in class could have brought it even closer to a real inquiry, and one that for second language learners going over the "concepts" could have improved the experience. One teacher mentioned that some students had "missed" real equipment, and one that it had not worked out well.

4 Discussion and Conclusion

On the whole, experiences from the teaching experiment were positive, and according to the reports of the teachers, the implementation of the simulation in the classroom had encouraging effects on both teachers and students in most cases. It was interesting and positive to see that teachers were mentioning many of the arguments that have been put forward in the literature. Almost all teachers referred to the learner-centric nature of the learning environment. Teachers also explicitly referred to affordances of the simulation (e.g., being clearer than real equipment and getting clear and quick measurement results) (e.g., [21]). It was also clear from the experiences of most of the teachers that they appreciated the effect that the learning environment had on their students' interest and engagement (cf. [12]). Many explicitly mentioned the fact that the learning environment was motivating and engaging and that the students liked being able to explore with the simulation in their answers on more than one occasion. Learning outcomes were not explicitly measured in this study, and in fact, some of the teachers had some doubts whether sufficient understanding had been obtained by all students or would transfer to understanding of circuits in the real world. As discussed earlier, learning outcomes in our earlier experimental studies suggest they do, but it would of course be good to affirm that also in a real classroom setting in future studies.

One of the generally positive things that came up through the comments of the teachers is that most of them clearly recognized that simulation-based inquiry learning emphasizes different aspects of learning (cf. [1]). Teachers for instance appreciated that they saw some students collaborating really well, but also that some others were not able to collaborate fruitfully, seeing this as something to develop. As mentioned before, teachers emphasized the engagement of their students and the fact that students were daring and being able to try out and test things, but some teachers also expressed concern that maybe not everyone learned to the same extent, for instance because, as one teacher mentioned, some students were progressing through the worksheets without proper understanding of the material they covered. It is interesting that these concerns were raised here (they have also been raised in relation the online teaching because of COVID-19) while that is something that is not often discussed in the context of regular instruction, even though it is clear that differences between pupils' performance also exist in that context. Therefore, an interesting question for future research is if that is because these differences are indeed more pronounced in these kinds of inquiry learning settings when learners need to take more control of and active role in their learning, or because teachers' pay more attention to these kinds of things when they try out new things (although all teachers involved in the study reported having previous experiences and perceptions of inquiry learning, it became clear during the training that the teachers generally used and were more comfortable with more teacher-centric methods in their teaching).

The observations and suggestions for improvement of the learning environment and teachers' own implementation show that most teachers in this study are in principle able to make material their own and adapt it to the needs of their classroom. This is positive and important, because one of the clear messages from this study is that it is virtually impossible to design a learning environment that can be used out of box or is perceived the same by all teachers; the same learning environment that was seen as very clear and working flawlessly by some teachers was assessed more critically by others. In addition, teachers also indicated different learning goals for their implementation: some emphasizing the process, others the knowledge outcomes and some both. This emphasizes the important role of teachers in relation to implementation of simulations in the classroom, or any other addition to the regular classroom repertoire. It requires their involvement, their understanding of the affordances and the pedagogical goals that they try to achieve. This is not trivial (see also Papaevripidou, Xenofontos, Hovardas, and Zacharia in this book) as was also illustrated by the expectations that teachers had beforehand. These clearly indicated that it is not always easy for teachers to estimate how things will go even when it concerns the students in their own class.

Of course teachers also mentioned things that could have been improved, and some of these would even be fairly easy to achieve. However, in light of the discussion above, rather than trying to incorporate all possible suggestions for improvement, it may be more important to support teachers' own agency in adapting the learning material to their own classroom [22]. In that case, it may be enough to provide teachers with learning material that can itself provide a positive experience, and the results of this study showed that for the majority of the teachers implementing this simulation-based learning environment had indeed been a positive experience. In combination with their recognition of the potential of these kinds of interactive environments in their classroom and explicit intention to use this kind of material more often, the fact that they saw opportunities for improvement could be seen as a positive outcome. The more complicated question is of course how to engage teachers in training and implementation experiences. One potentially fruitful approach would be to create positive experiences during the teacher training. The Penn and Ramnarain chapter in this book presents an example of the positive effects of such experience. Combining these experiences with an inquiry model as was done in this study may also be important, as it provides teachers with an interpretation (and evaluation) framework, and as one teacher indicated that introducing such framework to students may also be a useful addition to their learning experience.

Acknowledgements This work was supported by Grant no. 266189 from EU, in the context of the Ark of Inquiry project.

References

1. de Jong, T.: Moving towards engaged learning in STEM domains; there is no simple answer, but clearly a road ahead. J. Comput. Assist. Learn. **35**(2), 153–167 (2019). https://doi.org/10.1111/jcal.12337
2. Rocard, M., Csermely, P., Jorde, D., Lenzen, D., Walberg-Henrikson, H., Hemmo, V.: Science Education Now: A New Pedagogy for the Future of Europe. European Commission: Directorate-General for Research (2007). https://www.eesc.europa.eu/en/documents/rocard-report-science-education-now-new-pedagogy-future-europe
3. Hake, R.R.: Interactive-engagement versus traditional methods: a six-thousand-student survey of mechanics test data for introductory physics courses. Am. J. Phys. **66**(1), 64–74 (1998). https://doi.org/10.1119/1.18809
4. Bransford, J., Brown, A.L., Cocking, R.R. (eds.): How People Learn: Brain, Mind, Experience, and School (Expanded Edition). National Academy Press (2000)
5. Hazelkorn, E., Ryan, C., Beernaert, Y., Constantinou, C.P., Deca, L., Grangeat, M., Karikorpi, M., Lazoudis, A., Casulleras, R.P., Welzel, M., Kommission, E., Kommission, E., (eds.): Science Education for Responsible Citizenship: Report to the European Commission of the Expert Group on Science Education. Publications Office of the European Union (2015)
6. National Research Council: Next Generation Science Standards: For States, By States. Washington, DC: The National Academies Press (2013).https://doi.org/10.17226/18290
7. Klahr, D., Triona, L.M., Williams, C.: Hands on what? The relative effectiveness of physical versus virtual materials in an engineering design project by middle school children. J. Res. Sci. Teach. **44**(1), 183–203 (2007). https://doi.org/10.1002/tea.20152
8. Triona, L.M., Klahr, D.: Point and click or grab and heft: comparing the influence of physical and virtual instructional materials on elementary school students' ability to design experiments. Cogn. Instr. **21**(2), 149–173 (2003). https://doi.org/10.1207/S1532690XCI2102_02
9. Chang, K.-E., Chen, Y.-L., Lin, H.-Y., Sung, Y.-T.: Effects of learning support in simulation-based physics learning. Comput. Educ. **51**(4), 1486–1498 (2008). https://doi.org/10.1016/j.compedu.2008.01.007
10. Finkelstein, N.D., Adams, W.K., Keller, C.J., Kohl, P.B., Perkins, K.K., Podolefsky, N.S., Reid, S., LeMaster, R.: When learning about the real world is better done virtually: a study of substituting computer simulations for laboratory equipment. Phys. Rev. Spec. Top. Phys. Educ. Res. **1**(1), 010103 (2005). https://doi.org/10.1103/PhysRevSTPER.1.010103
11. Jaakkola, T., Nurmi, S.: Fostering elementary school students' understanding of simple electricity by combining simulation and laboratory activities. J. Comput. Assist. Learn. **24**(4), 271–283 (2008). https://doi.org/10.1111/j.1365-2729.2007.00259.x
12. Tapola, A., Jaakkola, T., Niemivirta, M.: The influence of achievement goal orientations and task concreteness on situational interest. J. Exp. Educ. **82**(4), 455–479 (2014). https://doi.org/10.1080/00220973.2013.813370
13. Steinberg, R.N.: Computers in teaching science: to simulate or not to simulate? Am. J. Phys. **68**(S1), S37–S41 (2000). https://doi.org/10.1119/1.19517
14. Pedaste, M., Mäeots, M., Siiman, L.A., de Jong, T., van Riesen, S.A., Kamp, E.T., Manoli, C. C., Zacharia, Z.C., Tsourlidaki, E.: Phases of inquiry-based learning: definitions and the inquiry cycle. Educ. Res. Rev. **14**, 47–61 (2015)
15. Jaakkola, T., Nurmi, S., Veermans, K.: A comparison of students' conceptual understanding of electric circuits in simulation only and simulation-laboratory contexts. J. Res. Sci. Teach. **48**(1), 71–93 (2011). https://doi.org/10.1002/tea.20386
16. Jaakkola, T., Veermans, K.: Effects of abstract and concrete simulation elements on science learning. J. Comput.-Assist. Learn. **31**(4), 300–313 (2015). https://doi.org/10.1111/jcal.12089
17. Jaakkola, T., Veermans, K.: Exploring the effects of concreteness fading across grades in elementary school science education. Instr. Sci. **46**(2), 185–207 (2018). https://doi.org/10.1007/s11251-017-9428-y

18. Jaakkola, T., Veermans, K.: Learning electric circuit principles in a simulation environment with a single representation versus "concreteness fading" through multiple representations. Comput. Educ. **148**, 103811 (2020). https://doi.org/10.1016/j.compedu.2020.103811
19. Veermans, K., Jaakkola, T.: Pedagogy in educational simulations and games. In: Cai Y., van Joolingen, W., Walker Z. (eds.) VR, Simulations and Serious Games for Education. Gaming Media and Social Effects, pp. 5–14. Springer, Singapore (2019). https://doi.org/10.1007/978-981-13-2844-2_2
20. Elo, S., Kyngas, H.: The qualitative content analysis process. J. Adv. Nurs. **62**, 107–115 (2008)
21. Wieman, C.E., Adams, W.K., Perkins, K.K.: PHYSICS: PhET: Simulations that enhance learning. Science **322**(5902), 682–683 (2008). https://doi.org/10.1126/science.1161948
22. Ketelaar, E., Beijaard, D., Boshuizen, H.P., Den Brok, P.J.: Teachers' positioning towards an educational innovation in the light of ownership, sense-making and agency. Teach. Teach. Educ. **28**(2), 273–282 (2012). https://doi.org/10.1016/j.tate.2011.10.004

Developing and Integrating an Augmented Reality App for Teaching and Learning About Enzymes

Wouter van Joolingen, Sui Lin Goei, Henri Matimba, Ryan Kyaw Thu Aung Ba, and Teresa Pedro Gomez Dias

Abstract In this chapter, we report on the development of an augmented reality (AR) app for teaching the working of enzymes. The aim of the study presented was to investigate how teachers and developers can work together in creating technology for education, in this case AR for teaching biology. In the project, four teachers from two secondary schools participated in a Lesson Study (LS) team with educators and developers of teacher training institutes and a professionals specializing in creating applications for virtual and augmented reality. We report on the process of development, the resulting app and lessons as well as on the first experiences in the classroom. The main conclusion on the process concerns the integration between lesson design and app development. Basic preparations on acquainting teachers with the use of AR and its content need to be made before starting the design and development cycle. Regarding the app itself, we identified improvements on the way AR can be a more integrated part of the learning activities.

Keywords Augmented reality · Inquiry-based learning · Lesson Study · Enzymes · Biology education

W. van Joolingen (✉) · H. Matimba
Freudenthal Institute, Utrecht University, Utrecht, The Netherlands
e-mail: w.r.vanjoolingen@uu.nl

S. L. Goei
LEARN! Learning Sciences, VU Amsterdam, Amsterdam, The Netherlands

S. L. Goei · T. Pedro Gomez Dias
Department of Human Movement and Education, Windesheim University of Applied Sciences, Zwolle, The Netherlands

R. K. T. A. Ba
Nanyang Technological University, Singapore, Singapore

Y. Cai et al. (eds.), *Virtual and Augmented Reality, Simulation and Serious Games for Education*, Gaming Media and Social Effects,
https://doi.org/10.1007/978-981-16-1361-6_11

1 Introduction

Creating technology-based environments that allow students to experience environments that they can explore and play around with have been built and studied over at least the last four decades [4, 24]. Examples are the experiences where a learner or student is fully immersed in a computer-generated virtual environment [3, 27], but also through the use of simulations whereby physical properties can be explained and taught [5, 16, 17, 22]. The technology can also be used in helping a learner train practical skills and so reduce hazardous situations or to make the training program more cost effective [23]. In short, through the coexistence of virtual and real environments, learners can experience phenomena that would otherwise be unattainable in the real world.

In this chapter, we focus on a study on the use of augmented reality in biology education for upper secondary school. As a science, biology stands out as complex. It studies life, in itself a very complex object of study, it also studies life at many levels, from molecular processes within the cell to complete ecosystems. As a result, biological science operates at many levels, and it is important to investigate the relations between levels. For instance, the processes on the level of individual organism (e.g., taking in food) may lead to emerging phenomena (e.g., food shortage) on the population level. This process can, on the other hand, be caused by processes on a smaller level of aggregation, (e.g., low blood sugar levels leading to hormone production that causes an urge to eat). In order to fully understand biological processes and systems, it is therefore required that reasoning takes place on different levels of aggregation and between these levels. Such reasoning is called, with an interesting visual analogy, *yoyo-ing* [13]. Yoyo-ing requires awareness of the existing structure of biological levels of aggregation as well as knowledge of the links between these levels.

Yoyo-ing is an important part of *systems thinking* [10–12, 25], which is an approach that applies generic system concepts such as systems' boundaries, their hierarchical structure, their components and relation between them to biological systems at all levels. Systems' thinking implies a systematic way of approaching a biological system, accounting for all aspects of studied biological phenomena.

Systems' thinking relies to a large extent on *models*. Models are used as a means to mediate between a person's ability to understand complexity and the original system [15, 18]. As an example, a scale model of a torso with organs can serve to deal with the complexity of the human body and the way organs are positioned and connected. The torso allows to inspect the relative position of the organs, see the way they are connected by blood vessels but at the same time the torso leaves out much detail that may be unnecessary for the purpose of the model. In such a way, the torso helps students and teachers to reduce complexity and focus on the relevant detail of a biological process.

Combining systems' thinking and models implies that models can exist on multiple levels of organization. For instance, based on the torso model, if we are interested in the workings of one specific organ, e.g., the stomach, we need a model that will reveal more detail of the stomach and the processes that take place within [2].

Augmented and virtual reality are very suitable applications for providing models at these multiple levels. Augmented reality allows to add information to the observed reality, e.g., labeling body parts or visualizing the inside of the body. Virtual reality can help to zoom in or out between layers, e.g., from the organism level, to the organ level all the way down to the level of molecular processes in the organs.

The purpose of the current study was to explore the potential of AVR in biology education by focusing on the use of AR to allow students to travel through the several layers involved in understanding digestion. The main focus was on the digestion of carbohydrates, in particular, sugars and starch, in chemical terms mono-, di- and polysaccharides. In the digestive system, the longer chains of saccharides are broken down by enzymes into smaller chains and eventually glucose molecules, which can be admitted to the bloodstream.

We developed and evaluated an instructional app that was built to test the advantage of learning about the concept of enzyme operation, for the context of carbohydrate digestion, with the aid of a handheld device. The idea behind the choice is that for the learning of the concepts involved, students would be able to mentally visualize the conceptual ideas concerning the learning objective. We chose for handheld devices (i.e., android phones and tablets) because this does not necessitate the use of systems where the user has to be fully immersed in the learning environment and that students can use the learning tool (app) in conjunction with more traditional learning methods. The latter gives the opportunity to cross reference learning effects with or without the aid of additional AR learning tools. Furthermore, it is interesting to test to what extent is it possible to incorporate deep mental learning or model thinking in a task using AR capabilities. The latter is one of the key objectives of twenty-first-century skills and learning [26].

1.1 Using a Lesson Study Approach to Design, Develop and Test Instructional Learning Tools

A major decision in the development of the app was that its development process was integrated with the design of the lesson in which it was planned to use. This means that the development of the lesson planned informed the development of the app. At least in theory, then the app could be tailored to the demands imposed by the lesson design and reversely, the lesson can be based on the possibilities the app offers.

In designing the lesson, we followed the approach of *Lesson Study* (LS) [6, 9, 14]. In this approach, a small team of teachers designs a *research lesson* around a research question. The lesson design is primarily focused on understanding and predicting student learning processes in response to the interventions in the lesson. Once a lesson is designed, it will be performed live by one of the team members where the other team members will observe student learning. In the Dutch model of LS, the focus of observation is on selected *case students* for whom detailed predictions will be made as part of the lesson plan, following the model of LS introduced by Dudley [8] and De Vries and colleagues [6]. After teaching and observing the lesson, it will be revised and taught a second time. In a reflection stage, the LS team will use the data collected from the whole cycle and potentially move on to a new or updated research question after which the cycle can repeat. Figure 1 displays the full LS cycle based on its representation by Stepanek and colleagues [19].

Designing research lessons entails knowledge about the learner, as well as the subject or topic that is being instructed and available teaching methods and approaches. A strong emphasis is made in understanding what the student needs to be able to learn or fulfill the learning task. In order to obtain this goal, the teacher scrutinizes his/her action in a team of peers concerning a chosen lesson or part of thereof.

In the current case, the development of the app was integrated within the Lesson Study cycle. This means that in early meetings, the team developed ideas for the app that were taken up by the developer. In principle, subsequent versions of the app should be discussed in various team meetings. We were interested in the way the LS process would interact with the design of the AR application, the way the app would be integrated in a lesson and how students would use the app to support

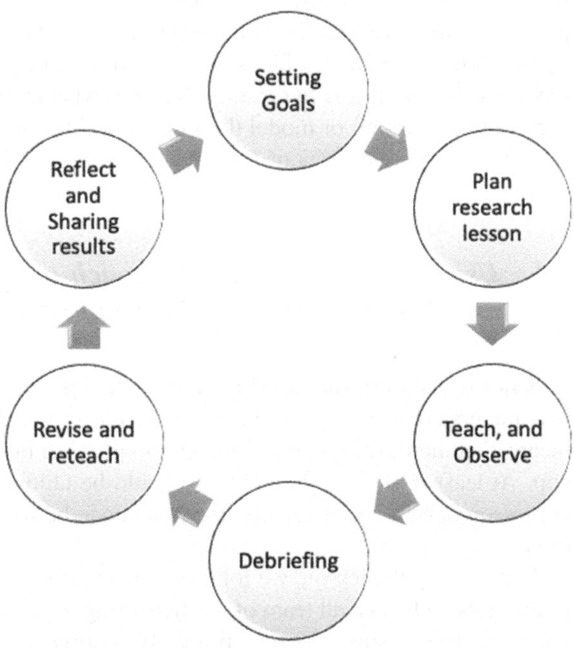

Fig. 1 Lesson Study cycle ref or based on the work by Stepanek and colleagues [19]

their learning. In the following, we describe the design phase of the app and the lesson as well as observations made in the lessons that were actually performed.

2 The Design Phase

2.1 *The Lesson Study Team and the Design Team*

The LS team consisted of four teachers of biology from two participating secondary schools, one LS expert who served as process facilitator, one teacher educator in chemistry, one teacher educator in biology and one researcher in science education who served as knowledgeable other with respect to modeling and ICT in education. The teachers had no or little experience with LS or with AVR. The team met several times over a period of a year to discuss the contents of the app, develop the actual lesson design, perform, observe the lesson and reflect on the process.

Next to the LS team, a smaller design team was set up, consisting of the teacher educator in chemistry and the researcher in science education, together with the app developer. This team was separate for two reasons. First, it was considered to be fruitful to separate the more technical talk from the talk on the lesson design. And second, the geographical separation (LS team in the Netherlands, developer in Singapore) made meetings with the full group hard to organize.

2.2 *Choice for Target Group*

In an early stage of the project, the team discussed the target group for the application. As the teachers in the group taught in different grades, it was decided to create an app that would be suitable for the use in both lower (grade 9) and upper (grade 11) secondary classes. This means that the app has to be functional for learners at the start of the learning about the domain, but also still be appealing to older students that have already acquired general knowledge concerning concepts such as enzymes and reaction conditions. This choice also implied that the app is not seen as a unit of learning in itself, but that it should be integrated in supporting materials and lesson designs, adapted to different target groups.

2.3 *Choice of a Model Enzymatic Reaction*

In the starting phase of this project, a choice had to be made by the participating secondary school teachers and university researchers which enzymatic reaction had to be chosen as the model reaction to be used in the app. Initially, the choice was made to build an app around theme of the occurrence lactose intolerance in humans. The occurrence of this intolerance is estimated to afflict about 65–75% of the global population [1], and it causes a condition characterized by symptoms such as stomach pain, bloating, gas and diarrhea, which are caused by lactose malabsorption. The rate of the occurrence also varies a great deal from region to region, from no more than 10% in Northern Europe to as much as 95% in some parts of Africa and Asia [7]. However, soon after the choice was made to build an app around the topic of lactose intolerance, it was abandoned, as it proved impossible to obtain the sought after structural chemical data needed to make an interesting and congruent scenario that could be used in the app. A major factor for this is the fact that the mechanistical considerations of the process are not yet fully understood in humans. After taking this into account, teachers and researchers decided to look for another model system.

This led first to the re-examination of the educational and methodological constraints governing the design of the scenario around the model system. A model system was sought that not only is specific toward a certain chemical process, but should also occur at different places in the human body, such as the muscles or the digestion system. This constraint narrows the choice of the possible model systems. It also gives rise to the opportunity to make allowances in the app to show the influence of vital secondary conditions, such as pH or temperature on the occurrence of the studied process.

The resulting model system chosen was the enzymatic reaction of the degradation of starch by amylase, following the description by Tomasik and Horton [20] as displayed in Fig. 2. This reaction plays an important and crucial role in the degradation of foods that contain starch, such as bread and potatoes. Starch consists of two types of molecules: One is a linear and helical chain (amylose), the other as a branched chain (amylopectin) build up from sugar units (saccharides). Plants and animals use these molecules to store energy. In our digestive system, starch is broken down into smaller fragments, called monosaccharides or glucose (1 unit), disaccharides (2 units) or polysaccharides (multiple units). This process occurs throughout different regions in the human digestive system. The process ultimately leads to the production of maltose (a disaccharide), which in turn is broken down to glucose in the intestines by the enzyme maltase. The enzyme that breaks down starch is called amylase, which occurs mainly in two variants, α- and β-amylase. α-amylase exists in saliva and can cut starch molecules in random places, resulting in polysaccharides of various lengths. β-amylase is produced in the pancreas and always splits off two glucose units at a time, resulting in maltose molecules. Important is that when swallowed, the α-amylase reaches the stomach and is *denaturized* meaning its molecular shape is unfolded due to stomach acids, and it loses its function.

Apart from starch also proteins, vitamins, cellulose, minerals and more substances are part of our diet. Within the context of this app, cellulose was decided to be particular interesting, as, just like starch, its molecules consist of chains or networks of glucose units, with as a difference that the way these units are bonded is

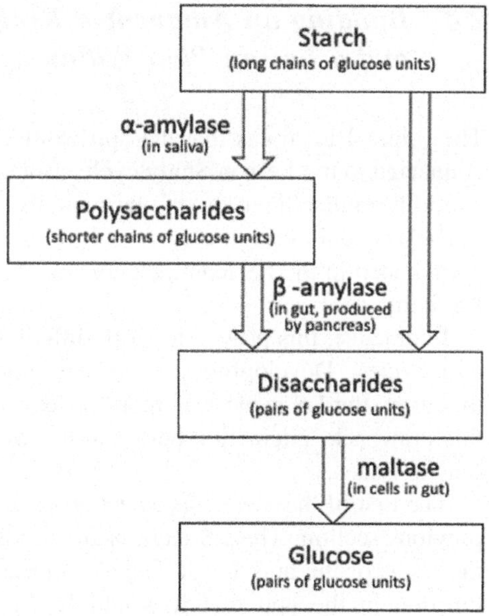

Fig. 2 Enzymatic conversion of starch [20]

different. As a result, amylase (or any other enzyme in the human digestive system) is not capable of digesting it, and it leaves the body undigested.

Taking these characteristics into account, the design team decided to create an app that focused on the digestive process of different kinds of food: bread, meat, fruit (a banana) and a soft drink. For each of the four items, the main components should be shown if where and how they would be digested by the two versions of amylase. These components could be starch and cellulose (bread), proteins (meat), fructose (a disaccharide) (banana) or maltose (soft drink). The app should allow students to choose one of these items and follow its digestion through a person's digestive system.

The app then could be used for several teaching goals as follows:

- To show the structure of the human digestive system
- To show that starch is digested (cut into smaller units)
- To let the student investigate the working of the two kinds of amylase
- To let the student explore three levels of biological organization (organism, organ and molecular) and move up and down between them.

An app that would provide views allowing for these tasks could be used both in lessons that support a first exploration of the digestive system, as well as a deeper understanding of the molecular processes that underlie the processing of starch in the body into glucose that can be dissolved in the bloodstream.

2.4 Building an Augmented Reality Application and a Lesson Plan Within a LS Cycle

The original idea was that the app would be designed in a separate design cycle, connected to the Lesson Study cycle, as depicted in Fig. 3. In the LS cycle, teachers would formulate their lesson plans and the resulting requirements for the app. Then in the secondary design cycle, the developer, together with the two experts who also participated in the LS team, versions of the app would be created for review by the LS team.

In practice, this proved to work differently, due to the different timescales of the two cycles. Developing the needed graphics took longer than expected, and therefore, the LS cycle was more or less halted for a relatively long time. Despite this drawback, interaction between the two development cycles took place on two crucial issues.

The first of these was the selection of the enzymatic reaction, as described in the previous section. The LS team at first chose for a different enzyme (lactase) to be the focus of the app, but it actually appeared from the first attempts to model this reaction in the app that it would be impossible to visualize. This initiated an exchange between the two cycles and the search for a better example enzyme reaction, resulting in the latter choice.

The second major decision that was made based on the interaction between the LS team was on the way the visualization would be integrated into the lesson. Here the lesson design steered the app development, in the sense that the scenarios were

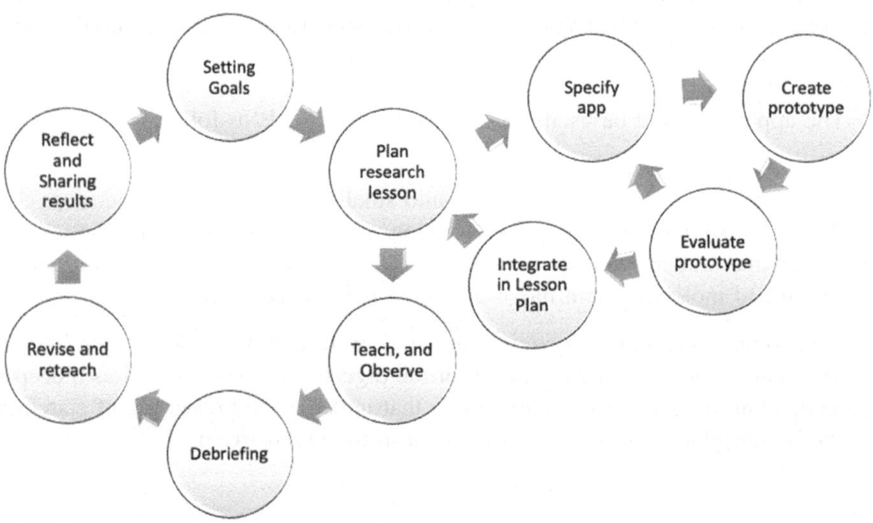

Fig. 3 Design cycle as an addition to the LS cycle

designed in the team, as part of the way the team foresaw students to explore the working of the enzymes, crossing the various levels of biological aggregation.

In the construction of the app, simple storylines were chosen all following the same principle. The learner starts at the macroscopic level, where a food is digested and subsequently followed as it takes its course through the human body. There are foods that contain starch, fibers, protein or a combination starch and fibers. After the digestion, the app takes the learner a level lower to the microscopic level of the molecules present in the food. It is possible to induce specific cleavage patterns by the enzymes (α- and β-amylase) when starch is present. This is governed by the position and type of enzyme in the body, so in the mouth you will only find the enzyme α-amylase and after the pancreas the enzyme β-amylase.

As a result, the app started from scanning a picture of a woman standing behind a table filled with food items (see Fig. 4). The app provides an overlay over the picture of a skeleton showing the relevant organs of the digestive system. Student can then pick a food item which takes them to a screen displaying the major organs of the digestive system: mouth, stomach, gut and bowel. Students can zoom in further on these organs and inspect the stage of digestion in that organ: α-amylase cutting up the starch in the mouth, its denaturation in the stomach, β-amylase chipping off disaccharides in the gut and the resulting glucose entering the bloodstream in the gut and the bowel. Moreover, depending on the food chosen, the presence of starch and other components of the food taken in are shown as well as the effect the both versions of amylase have on them (or not).

Lessons were designed for two levels of secondary education. For the 11th grade (age 16–17), who, in earlier lessons were taught the basic working of enzymes as well as the basic component, the students were to start with group work to investigate the working of the two types of amylase *before* interacting with the app. For this, they could use a reference book for the sciences to study their molecular structure and their working. Also, they used a depiction of the human digestive channel to label its components, as a reminder of the structure of the digestive system and to locate the place where the various stages of digestion take place. After this, they were asked to check the digestive path of starch, fibers and proteins using the app and to collect the results in a table, matching food components to enzymes and organs. The table was then made subject in a class-wide discussion, in which the teacher filled in the table, based on contributions by the separate groups.

In the design for ninth grade (age 14–15) classes, the app was used in a more explorative manner. The same table as for the 11th grade was used, but students were instructed to use the app from the beginning to investigate in which organs the enzymes operated. They used the information in the app to fill in the same table as was used for the 11th graders, and also this table was discussed with the whole class. This discussion was followed by an explanation by the teacher on how enzymes work, using a large display of the molecular structure of amylase.

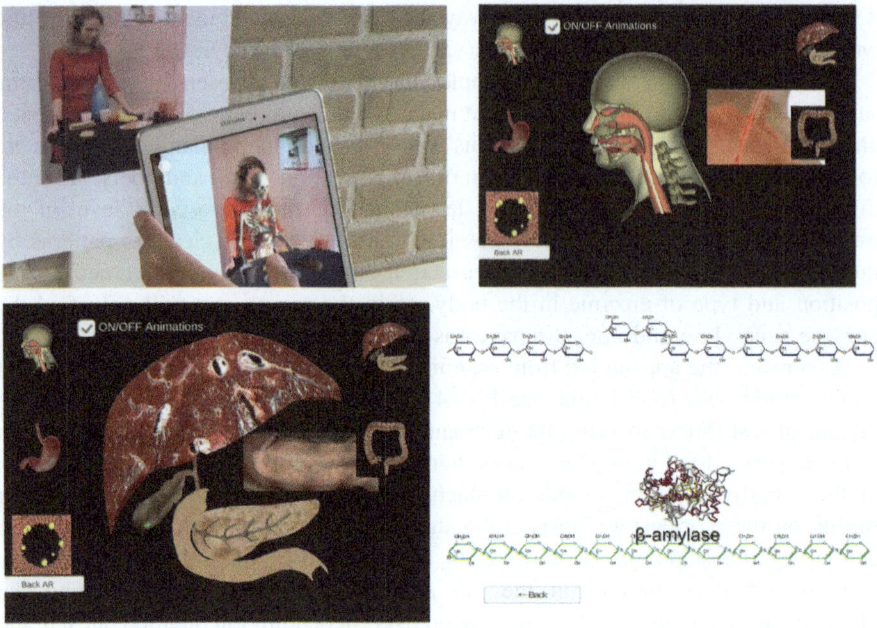

Fig. 4 Four screenshots of the app. The top left picture shows the app while the trigger picture is being scanned on the macroscopic level. Pictures in top right and bottom left show two aspects of the organ level (mouth and gut, with associated organs). The bottom right picture shows the molecular level and allows the student to use the amylase to cut the starch chain on top, but not the cellulose (fiber) chain on the bottom

2.5 *The Lessons*

The lessons were deployed as planned, once in a secondary school in the east of the Netherlands, where one of the teachers of the LS team performed the lesson for 11th grade and two times in a 9th grade class in the center of the Netherlands. In each class, three *case students* were selected who were observed by members of the LS team, and all research lessons were video recorded. Case students were also interviewed after the lesson. Students' active informed consent was gathered by their teachers, and in the case when students were underage, active informed consent of their parents was obtained.

In the 11th grade class, the lesson was performed as planned. Notably, students refrained from using the app until they completed their investigation in the specific enzymes and performed the learning tasks as instructed. After the investigation, students took the tablet and scanned the picture. Operation of the app proved no problem, after some experimenting, all students could operate the features of the interface. One of the case students enjoyed the "cutting operation", in which the enzyme can be used to cut the chains of starch into disaccharides.

One thing that was very clear in the lesson is that students typically used the app in a very systematic way: scan the picture, choose a food item and work systematically through the various organs to check whether amylase works on the substances that are present in the food. In the post-lesson interviews, case students stated that they liked the app and were satisfied with the lesson.

Also, in the ninth grade classes, the lesson went as planned. Here it became clear from students behavior as well as from the interviews with the case students that the students perceived the app as a game. One of the case students explicitly used this word in the way he described and analyzed the app. From this perspective, he critiqued the app on its game features and the way it could be improved.

3 Evaluation and Discussion

In this chapter, we described the development of an augmented reality app that provides insight into the working of amylase as part of the human digestive system. The app was designed in a collaboration between teachers and a design team, integrated in two lessons designs, which were tested within the classroom in three lessons. The cycle of design followed the LS approach, connected to the development team that implemented the app. The resulting lessons in which the app was employed were then performed and observed. In this section, we summarize the main findings from both the design phase and the performance of the lessons.

In the design phase, the initial intention was to integrate the design and development cycles, as in a rapid prototyping approaches [21]. The idea was that this would lead to a kind of agile development in which the LS team would be able to review subsequent versions of the app, decide on how these could be integrated in a lesson leading to more refined specifications. In practice, however, the process did not unfold as planned. Two factors seem important to this.

The first factor is that at the start of the cycle many decisions had to be taken which took a lot of investigation. For instance, although the plan was to focus on an enzyme, it took time to determine which enzyme was the most appropriate, both from an educational perspective and from a technical perspective. It was not helpful to this process that for the teachers this was the first time they were working with an AR environment, so they lacked experience and knowledge on its possibilities and limitations. As a consequence, it was hard to derive concrete guidelines for the app developer.

The second factor was that the developer was located in Singapore, while the team was in the Netherlands. Due to time differences and teaching schedules, a meeting of both teams together was impossible so two team members served as a liaison between teachers and developer. This indirect communication was probably not helpful for the speed of development. The resulting fact that the app development was slow relative to the LS cycle effectively stalled the LS cycle for a considerable time during which a version of the app was developed on partial specifications. Time constraints prevented that the design—evaluate—redesign cycle for the app could not be fully exploited after this time.

A lesson to be learnt from this is that it is important to prepare the LS and development cycle before they actually start. In the current case, this would mean that choices for an enzyme would need to be investigated and explored before the actual LS and development cycles would have started. Also, initial development of example displays, if only as a mock-up, could have boosted the creativity of the teachers for the lesson design. This would allow the team to use the basic material to discuss flows of the lesson, exploring the different ways the app could have been used in the lesson.

With respect to the app itself, the two lesson designs, albeit very similar to each other, showed that it was possible to use the app in both an explorative and confirmatory way. However, the way the students used the app showed that the paths they followed were quite linear: scan the picture, choose food and check the processes within the organs. While this did a good job in linking the various levels of aggregation visually, it left little room for active exploration. To this end, the app should be extended with more enzymes and food components so that students can explore all the combinations. The current version of the app also lacks a visualization of the actual process of enzyme docking, which would be helpful in understanding *why* a certain enzyme can break down a molecule.

The same linear way of working with the app also raises the question on how essential the augmented reality part is of the app design. In effect, the picture was only used to bring up the start screen of the sequence, where the user can choose a food item. In order to stress the relation with the macroscopic level, it could be considered to require the picture to stay in view and use animations to explicitly stress the movements between levels of biological aggregation. Then the AR aspect of the app could be more prominent and useful from an educational perspective.

Concluding we can state that the development process and the resulting app on enzymes provided valuable insights into the way teachers can be involved in the design of advanced technological tools in the classroom. While the process did not always turn out as planned, we can learn from the experience. The main lesson learnt is the importance of preparation, the availability of good examples and basic material is essential to get a proper development going in which teachers and developers collaborate in an effective way.

Acknowledgements The authors would like to thank Hilde Rinia and Bep Portengen of Goois Lyceum in Bussum and Jan Jaap Wietsma and Corine Hijmissen of Greijdanus College in Zwolle (both in the Netherlands) for their participation in the Lesson Study team. This project was funded by SIA KIEM, project number 107164.

References

1. Bayless, T.M., Brown, E., Paige, D.M.: Lactase non-persistence and lactose intolerance. Curr. Gastroenterol. Rep. (2017). https://doi.org/10.1007/s11894-017-0558-9
2. Bierema, A.M.K., Schwarz, C.V., Stoltzfus, J.R.: Engaging undergraduate biology students in scientific modeling: analysis of group interactions, sense-making, and justification. CBE Life Sci. Educ. (2017). https://doi.org/10.1187/cbe.17-01-0023
3. Cai, Y., Tay, C.T, Ngo, B.K.: Introduction to 3D immersive and interactive learning. In: 3D Immersive and Interactive Learning, vol. 9789814021906, pp. 1–16. Springer, Singapore (2013). https://doi.org/10.1007/978-981-4021-90-6_1
4. De Jong, T.: Technological advances in inquiry learning. Science **312**(5773), 532–533. https://a.j.m.dejong@utwente.nl (2006)
5. De Jong, T., Van Joolingen, W.R.: Scientific discovery learning with computer simulations of conceptual domains. Rev. Educ. Res. **68**(2) (1998)
6. de Vries, S., Verhoef, N., Goei, S.L.: Lesson Study: een praktische gids voor het onderwijs. Garant (2016)
7. Deng, Y., Misselwitz, B., Dai, N., Fox, M.: Lactose intolerance in adults: Biological mechanism and dietary management. Nutrients **7**(9), 8020–8035 (2015). https://doi.org/10.3390/nu7095380
8. Dudley, P.: Lesson Study. Professional Learning for our Time (2015)
9. Dudley, P.: Teacher learning in lesson study: what interaction-level discourse analysis revealed about how teachers utilised imagination, tacit knowledge of teaching and fresh evidence of pupils learning, to develop practice knowledge and so enhance their pupils' learning. Teach. Teach. Educ. **34**, 107–121 (2013). https://doi.org/10.1016/j.tate.2013.04.006
10. Gilissen, M.G.R., Knippels, M.-C.P.J., van Joolingen, W.R.: Bringing systems thinking into the classroom. Int. J. Sci. Educ. 1–28 (2020). https://doi.org/10.1080/09500693.2020.1755741
11. Gilissen, M.G.R., Knippels, M.C.P.J., Verhoeff, R.P., van Joolingen, W.R.: Teachers' and educators' perspectives on systems thinking and its implementation in Dutch biology education. J. Biol. Educ. 1–12 (2019). https://doi.org/10.1080/00219266.2019.1609564
12. Klopfer, E., Yoon, S., Um, T.: Teaching complex dynamic systems to young students with starLogo. J. Comput. Math. Sci Teach. **24**(2), 157–178 (2005). papers3://publication/uuid/5118CE4F-B0F6-4568-ABCF-9B025E19208B
13. Knippels, M.-C.P.J., Waarlo, A.J.: Development, uptake, and wider applicability of the yo-yo strategy in biology education research: a reappraisal. Educ. Sci. **8**(3), 129 (2018). https://doi.org/10.3390/educsci8030129
14. Lewis, C.C., Perry, R., Murata, A.: How should research contribute to instructional improvement? The case of lesson study. Educ. Res. **35**(3), 3–14 (2006). https://doi.org/10.3102/0013189X035003003
15. Magnani, L., Nersessian, N., Thagard, P.: Model-Based Reasoning in Scientific Discovery. Kluwer (2012)
16. Rutten, N., van der Veen, J.T., van Joolingen, W.: Inquiry-based whole-class teaching with computer simulations in physics. Int. J. Sci. Educ. 1–21 (2015). https://doi.org/10.1080/09500693.2015.1029033
17. Rutten, N., van Joolingen, W., van der Veen, J.T.: The learning effects of computer simulations in science education. Comput. Educ. **58**(1), 136–153. https://linkinghub.elsevier.com/retrieve/pii/S0360131511001758 (2012)
18. Speth, E.B., Shaw, N., Momsen, J., Reinagel, A., Le, P., Taqieddin, R., Long, T.: Introductory biology students' conceptual models and explanations of the origin of variation. CBE Life Sci. Educ. **13**(3), 529–539 (2014). https://doi.org/10.1187/cbe.14-02-0020
19. Stepanek, J., Appel, G., Leong, M., Turner Mangan, M., Mitchell, M.: Leading Lesson Study: A Practical Guide for Teachers and Facilitators. Corwin Press (2007)

20. Tomasik, P., Horton, D.: Enzymatic conversions of starch. Adv. Carbohydr. Chem. Biochem. (2012). https://doi.org/10.1016/B978-0-12-396523-3.00001-4
21. Tripp, S.T., Bichelmeyer, B.: Rapid prototyping: an alternative instructional design strategy. Educ. Technol. Res. Dev. **38**, 31–44 (1990). papers3://publication/uuid/BF061BFA-9903-4080-AC5D-75A8310CAA28
22. van Joolingen, W., de Jong, T., Dimitrakopoulou, A.: Issues in computer supported inquiry learning in science. J. Comput. Assist. Learn. **23**(2), 111–119 (2007). https://doi.org/10.1111/j.1365-2729.2006.00216.x/full
23. van Joolingen, W., de Jong, T., Lazonder, A.W., Savelsbergh, E.R., Manlove, S.: Co-lab: research and development of an online learning environment for collaborative scientific discovery learning. Comput. Hum. Behav. **21**(4), 671–688. https://www.sciencedirect.com/science/article/pii/S074756320400189X (2005)
24. van Joolingen, W., Zacharia, Z.C.: Developments in inquiry learning BT-technology-enhanced learning. In: Technology-Enhanced Learning. Issue Chapter 2, pp. 21–37. Springer, Netherlands (2009). https://doi.org/10.1007/978-1-4020-9827-7_2
25. Verhoeff, R.P., Knippels, M.-C.P.J., Gilissen, M.G.R., Boersma, K.T.: The theoretical nature of systems thinking. Perspectives on systems thinking in biology education. Front. Educ. **3**, 40 (2018). https://doi.org/10.3389/feduc.2018.00040
26. Voogt, J., Roblin, N.P.: A comparative analysis of international frameworks for 21st century competences: implications for national curriculum policies. J. Curriculum Stud. **44**(3), 299–321 (2012). papers3://publication/uuid/CC8707B4-C128-4ED5-B290-C9FEC3E3AF72
27. Wu, H.K., Lee, S.W.Y., Chang, H.Y., Liang, J.C.: Current status, opportunities and challenges of augmented reality in education. Comput. Educ. (2013). https://doi.org/10.1016/j.compedu.2012.10.024

AR Simulation of Cardiac Circulation

Ryan Kyaw Thu Aung Ba, Lihui Huang, Siti Faatihah Binte Mohd Taib, Jieqiong Chen, Ye Kyaw Aung, and Yiyu Cai

Abstract Cardiac circulation is traditionally a difficult topic in biology for students from both high schools and colleges to learn due to its complexity and dynamic process. Cadavers can be used as a good tool for students to learn cardiovascular system. Unfortunately, this approach comes with some major concerns because of the collapsed anatomic structures and no blood flowing. Also, cadavers are normally available only in medical schools. Human subjects especially cardiac patients are ideal for students to learn human circulation from the perspectives of both structures and functions. This, however, is infeasible due to ethical and other constraints. As such most students learn cardiac circulation by reading text books, attending lectures, viewing images, and perhaps manipulating heart models. In this chapter, we will present our efforts in developing augmented reality (AR) technology to enhance the learning of cardiac circulation. More specifically, a book-based AR app is designed to allow each and every student to learn the cardiac structure and function by interactive playing.

Keywords Cardiac circulation · Augmented reality · Technology enhanced learning

1 Background

The coronary circulation system provides blood supply to the heart muscle. The coronary circulation begins near the origin of the aorta by two coronary arteries: the left coronary artery and the right coronary artery. After nourishing the heart muscle, blood returns through the coronary veins into the coronary sinus and from this one into the right atrium. Back flow of blood through its opening during atrial systole is

R. K. T. A. Ba · J. Chen · Y. K. Aung
VARTEL Network, Singapore, Singapore

L. Huang · S. F. B. M. Taib · Y. Cai (✉)
Nanyang Technological University, Singapore, Singapore
e-mail: myycai@ntu.edu.sg

Y. Cai et al. (eds.), *Virtual and Augmented Reality, Simulation and Serious Games for Education*, Gaming Media and Social Effects,
https://doi.org/10.1007/978-981-16-1361-6_12

prevented by the valve of the coronary sinus. The smallest cardiac veins drain directly into the heart chambers. The heart pumps oxygenated blood to the body and deoxygenated blood to the lungs. In the human heart, there is one atrium and one ventricle for each circulation, and with both a systemic and a pulmonary circulation there are four chambers in total: right atrium, left atrium, right ventricle, and left ventricle. The right atrium is the upper chamber of the right side of the heart. Blood that is returned to the right atrium is deoxygenated and passed into the right ventricle to be pumped through the pulmonary artery to the lungs for re-oxygenation and removal of carbon dioxide. The left atrium receives fresh oxygenated blood from the lungs through the pulmonary veins, which is passed into the strong left ventricle to be pumped through the aorta to different organs of the body.

Cardiovascular system is an important topic in biology education at secondary level or medical school. It is, however, difficult for students to learn the structures and functions due to its complex and dynamic nature. This paper presents our work on technology enhanced learning of the cardiac circulation. More specifically, augmented reality (AR) technology is developed to help students visualize cardiac circulation system. Students can use the AR tool created to interactively learn the anatomical structure of arteries and veins, valve, atriums, and ventricles as well as the functions of cardiac circulation.

The work reported here is part of our continual effort in understanding human cardiovascular or cardiac system with minimally invasive surgery or intervention [1, 2, 8–10]. Over the years, we have been investigating virtual reality (VR) technology for educational use [3–7]. In the following, we discuss the design and development of an AR tool for technology enhanced learning of cardiac circulation.

2 From Traditional to Technology Enhanced Learning

Cardiac circulation is traditionally a difficult biological topic for students from both high school and college to learn due to its complexity and dynamic process. Cadavers can be used as a good tool for students to learn cardiovascular system. Unfortunately, this approach comes with some major concerns because of the collapsed anatomic structures, and no blood flowing. Also, cadavers are normally available only in medical schools. Human subjects especially cardiac patients are ideal for students to learn human circulation from the perspectives of both structures and functions. This, however, is infeasible due to ethical and other constraints. As such most students learn cardiac circulation by reading text books, attending lectures, viewing images, and perhaps manipulating heart models.

In this project, we are keen to develop AR technology assisting students in their learning of cardiac circulation. With this AR tool developed, students when read the textbook will be able to view the cardiac system in 3D, manipulate the heart structure interactively, and simulate the heart function dynamically. The advantage of this technology enhanced learning is that it allows students to learn the topic in schools or after school at the pace that they are comfortable with. They can use low cost devices such as smartphones, iPhones, or iPADs to do their learning. This way, learning becomes visual and tangible similar to the use of cadavers. Furthermore, dynamic process of cardiac circulation will be available with the app tool, which makes learning of this difficult topic intuitive.

3 The Design and Development of the AR App

The AR tool is designed for secondary school students to learn the cardiac circulation. It is built for smartphones or iPhones. Its contents has two major parts (Fig. 1): structure and function. This interactive tool is designed for students to learn different types of anatomy of the cardiovascular system including coronary arteries, coronary veins, atrium, ventricles, valve, sinus, pumps, and so on, as well as the simulation of the blood flow, the flow direction, the open and close of valve, etc.

4 Result

Figure 2 shows the AR tool developed for the learning of cardiac circulation. Currently, the app is used together with the textbook for biology at secondary level, recommended by Singapore Ministry of Education. With the app installed in their smartphones, iPhones or IPADs, students can play with the app interactively to learn anatomic structure and function of the cardiac circulation system. The simple design of this app allows students to learn cardiac circulation easily through playing.

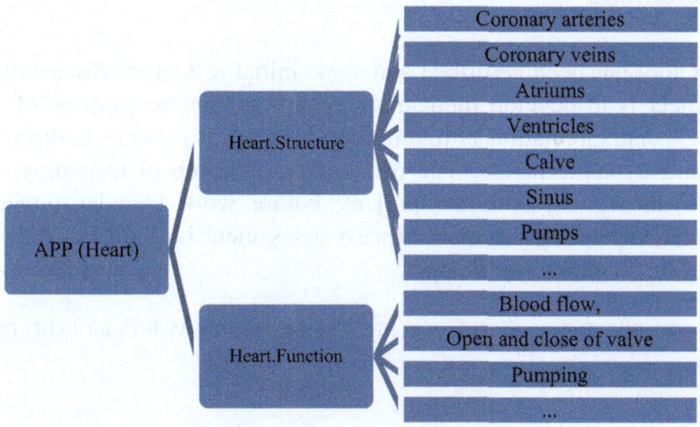

Fig. 1 Design of APP for heart circulatory

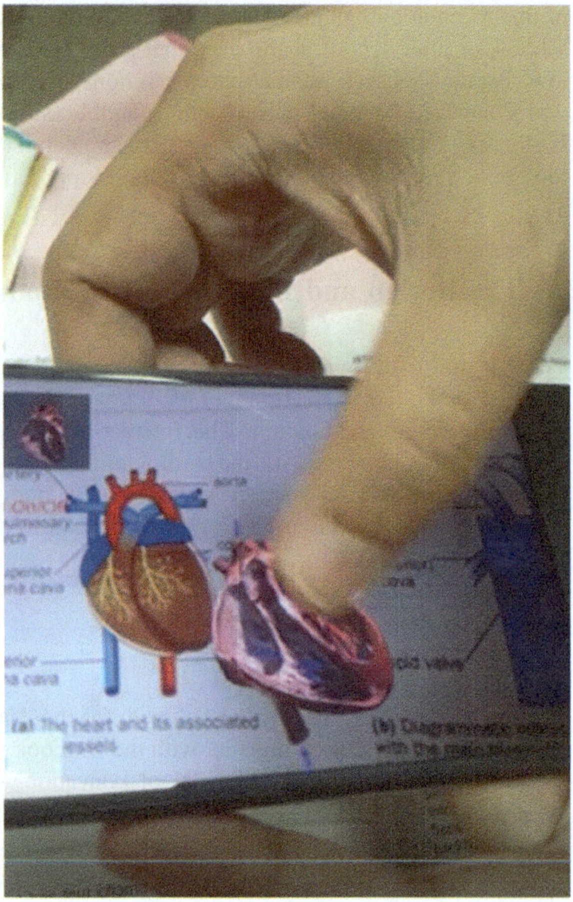

Fig. 2 Learning cardiac circulation using a smartphone

5 Conclusions

A new AR tool has been described and some initial results are discussed. The idea of this project is to develop innovative technology for the purpose of enhancing learning. Cardiac circulation is traditionally a difficult topic in biology due to its complex and dynamic nature. The pedagogical research of technology enhanced learning is undergoing with the project. Future work includes qualitative and quantitative analysis. Stealth or in process assessment is also of our interest to be integrated into the developed app.

Acknowledgements This research is partially supported by the MaF Grant (AFD 06/16 CY).

References

1. Cai, Y., Chui, C.K., Ye, X.Z., Fan, Z., Anderson, J.H.: Tactile VR for hand-eye coordination in simulated PTCA. Comput. Biol. Med. **36**(2), 167–180 (2006)
2. Cai, Y., Chui, C.K., Ye, X.Z., Anderson, J.H.: Hand-eye coordination in VR simulated revascularization. Int. J. Comput. Assist. Radiol. Surg. **1**(Suppl. 7), 255–257 (2006)
3. Cai, Y., Chia, K., Thalmann, D., Zheng, J.M., Kee, N., Thalmann, N.M.: Design and development of a virtual dolphinarium for children with autism. IEEE Trans. Neural Syst. Rehabil. Eng. **21**(2), 209–217 (2013)
4. Cai, Y., Chiew, R., Nay, Z.T., Indhumathi, C., Huang, L.: Design and development of VR learning environments for children with ASD. Interact. Learn. Environ. **25**(8), 1098–1109 (2017)
5. Cai, Y.: Introduction to 3D immersive and interactive learning. In: 3D Immersive and Interactive Learning, pp. 1–16. Springer, Singapore (2012)
6. Cai, Y., Goei, S.L., Trooster, W. (eds.): Simulation and Serious Games for Education. Springer, Singapore (2016)
7. Cai, Y., van Joolingen, W., Walker, Z. (eds.): VR, Simulation and Serious Games for Education. Springer, Singapore (2018)
8. Chiang, P., Zheng, J.M., Yu, Y., Mak, K.H., Chui, C.K., Cai, Y.: A VR simulator for intracardiac intervention. IEEE Comput. Graphics Appl. **33**(1), 44–57 (2013)
9. Chiang, P., Zheng, J.M., Mak, K.H., Thalmann, N.M., Cai, Y.: Progressive surface reconstruction for heart mapping procedure. Comput.-Aided Des. **44**, 289–299 (2012)
10. Wang, Y., Chui, C.K., Lim, H.L., Cai, Y., Mak, K.H.: Real-time interactive simulator for percutaneous coronary revascularization procedures. Comput. Aided Surg. **3**(5), 211–227 (1998)

360° Video for Immersive Learning Experiences in Science Education

Timothy Ter Ming Tan and Aik-Ling Tan

Abstract The availability of affordable, consumer-oriented video cameras capable of recording high-quality, full-motion video in 360° in every direction has enabled a rapid proliferation of what is commonly referred to as VR360 video across social and entertainment media, especially in travel, adventure sports and virtual location tours, and particularly as a trade tool for the real-estate industry. This chapter examines the uses, affordances, challenges, needs, and potential of VR360 and stereoscopic 180° (VR180) media for effective learning of science through the development and application of immersive learning experiences (ILE). ILEs are especially suited to support out-of-classroom (OOC) learning and/or where the use of first-person view (FPV) perspective may be beneficial, such as field trips and laboratory-based practical work. Essentially, these types of learning activities are viewed as important aspects of science education by engendering the embodied sense of "being there" and "doing". ILEs based on VR360 video have the potential to provide a sense of this embodied presence and engagement and thus can serve primarily as adjunct and supporting resources to traditional OOC and laboratory activities, and if necessary, replace these activities.

Keywords VR360 · Immersive reality · Instructional media · Science practical work · Out-of-classroom learning

1 Introduction

1.1 Background

The growing pervasiveness of augmented and virtual reality (AVR) technologies, proliferation of consumer devices that provide immersive experiences, the relative ease of AVR content creation, and the growing volume of AVR content accessible

T. T. M. Tan (✉) · A.-L. Tan
National Institute of Education, Nanyang Technological University, Singapore, Singapore
e-mail: timothy.tan@nie.edu.sg

Y. Cai et al. (eds.), *Virtual and Augmented Reality, Simulation and Serious Games for Education*, Gaming Media and Social Effects,
https://doi.org/10.1007/978-981-16-1361-6_13

to the general public create increasing expectations among consumers for easy and frequent access quality media of that nature. Students at all levels, or learners in general, similarly have ever higher exposure to AVR material and thus become increasingly accustomed to consuming such media resources outside of school. The uptake and use of such technologies within a school or educational context remain mixed and tend towards a conservative and cautious approach with limited deployment. Yet, the importance and applicability of AVR technology in education to create appropriate immersive learning experiences (ILE) are widely recognized [4, 16, 25]. In Singapore, for example, the Info-communications Media Development Authority (IMDA) considers education as one of the key areas for the applied use of immersive experiences [14] and immersive media as one of four frontier technologies that underpin Singapore's *Digital Economy* plan [20].

While AVR and immersive technologies are already considered "mainstream" [14, 17], the limited use in school settings is understandable and can be attributed to many factors. Among the obvious might be the need for capable computer devices and display systems such as head-mounted devices (HMD), potentially in substantial quantities for class use, the need for a certain level of media authoring expertise, and the time and labour needed to develop and create such resources. Perhaps less obviously, instructors may also struggle to find suitable use-cases or contexts in which to employ AVR, due in part to our limited knowledge of pedagogically appropriate ways in which to wield AVR in the classroom.

This chapter attempts to address aspects of this conundrum by discussing the ways in which low-cost consumer-grade hardware, readily accessible software tools and services, and a relatively easy-to-use subset of AVR modalities can be utilized directly by instructors working individually or in small in-house teams to develop useful and pedagogically sound learning materials. This particular focus on affordable consumer-oriented technologies and simple workflows is intended to encourage educators in and ease their adoption of VR360 as a pedagogical tool.

1.2 Immersive Learning Environments and VR360

Immersive learning experiences (ILE, or simply, *immersive learning*) situate learners in a simulated or mock environment that envelopes them and provides an overall illusion of engaging in the actual activities represented, for the purposes of learning [9, 24]. We prefer this original use of the abbreviation ILE over the term "immersive learning *environments*" which has become more common in recent years, because the former explicitly considers the learner's holistic *learning experience*, as opposed to the latter's implied focus on the created *environment*, since learning is derived from more than just the environment it occurs in. ILEs can be created within the real-world as, for example, a scenario-based role-play crime scene investigation [6], but perhaps more typically refer to virtual ILEs that employ some form of AVR simulation [9, 24].

Of particular interest to us are virtual ILEs that employ seamless, omnidirectional 360° images or video that can be viewed as if projected within a sphere surrounding the user, commonly referred to as VR360 media. A subset of this type of media allows for viewing within a 180° hemisphere, but feature stereoscopic imaging to convey the perception of visual depth and known as VR180. Unless specifically referring to VR180, we will use the term VR360 to collectively describe both forms. We distinguish VR360-based media or ILEs from other AVR technologies in that VR360 is primarily based on photographically or videographically *recorded* imagery from the real world as opposed to synthetically rendered visuals in VR, and in that VR360 does not have a *virtual world* or its associated 3D objects with which the user can manipulate. Where manipulatable VR objects do feature within the ILE, we would consider these to be hybrid AVR-VR360 platforms.

The media-sharing platform YouTube enabled the publishing of VR360 videos in 2015 [8] and VR180 videos in 2017 [3]. These events essentially marked the beginnings of the mass adoption of VR360 media across many sectors such as education, entertainment, advertising, tourism, and especially in the real-estate market. Interestingly, there appears to be a growing interest in the streaming of real-time VR360 for various purposes, and notably, VR360 has been found to have a positive impact among viewers of news reporting or *immersive journalism* [31, 34], where the conveying and retention of factual information are, similar to the field of education, of key importance.

Why and where, then, might ILEs featuring VR360 media be useful in education? In considering the potential application areas for VR360 media in science education, two key areas were identified as particularly suited to its use, and furthermore, these two areas have relatively minor but persistent challenges that could potentially be addressed by VR360 technology. These are the areas of out-of-classroom (OOC) learning activities and science laboratory practical work. The potential for VR360-based media to support the learning of, and overcome certain inherent challenges in, OOC and practical work activities by allowing individual learners equitable, consistent, and repeatable vantage points to view and learn from the phenomenon, event or procedure on show will be discussed in greater detail later. But principally, VR360 provides the learner with the ability to immersively experience these two forms of science instruction, from which experience we might expect the learner to become more familiar with the actions and procedures these science learning activities entail and indeed, be able to rehearse for in mental and/or embodied ways. Practice within virtual environments can be as effective or even more so than practice in real life due to the immediate real-time feedback the learner can receive from engaging in tasks, from which to undertake corrective action and hence learn from [10]. Furthermore, virtual environments can be controlled or set up to have specific conditions that might not be readily replicated in the real world, thus broadening the range of learning experiences available to the learner [24]. It is also known that VR360-based ILEs, when used for familiarization purposes to new experiences, can reduce anxiety [23] which would naturally be useful to improve the individual's readiness to learn.

1.3 Creating and Viewing VR360 Media

The development of *all-in-one* camera devices capable of natively capturing omnidirectional static images and video footage was arguably one of the key reasons for the rapid increase in popular interest in VR360 media in recent years. Prior to such integrated devices, only specialist and expensive equipment or the use of many conventional cameras mounted on cumbersome rigs were required to capture multiple images requiring software stitching of the individual images into the composite whole.

Purpose-made VR360 devices have since become widely available, affordable, compact, and consumer-oriented. These include VR360 cameras that capture still or video images either to memory storage for later download, and/or directly to connected mobile devices for immediate or even live viewing. Stand-alone HMDs that do not require a desktop class computer to function and that are capable of playing back various forms of AVR media, running AVR games, and so on have also become readily available to the consumer. These are essentially HMDs with a built-in mobile device class processor and, like most current generation mobile devices such as smartphones and tablets, have the processing capability to view AVR as well as high-quality full-motion VR360 video. A smartphone can also be converted into an ad hoc HMD using *Google Cardboard* or similar, while the flat-screen of a mobile device when used in a hand-held manner, is also an accessible way in which to view VR360 resources. The built-in accelerometers in most of such devices allow the phone or tablet screen to function as a *viewport* into the VR360 environment. The direction and orientation a user holds up the device present an image on the screen that makes the screen appear to be a "window" into the "surrounding" VR360 environment. A desktop or laptop computer screen can perform a similar function, except that the viewport or viewpoint shown in the image is slewed via keyboard or mouse control, rather than physically positioning the computer screen.

The capability, affordability, and consumer orientation of the devices needed for VR360 allow school teachers and other educators the means and ability to create their own VR360-based lesson resources and also enable their learners to easily view such resources in the classroom or from a remote location using a variety of personal devices. Detailed tutorials and methodological considerations for the design and production of VR360-based learning resources and virtual tours can be found in the literature for various contexts such as the classroom [7], the medical field [23], and cultural heritage education [5].

It should be clear at this point that the technology (hardware and software) and costs involved are within reach for most educators, schools, and institutions, and that its relative ease-of-use should enable rapid production of teaching materials by teachers themselves on an in-house basis, without the need for specialized technologists or artists. But beyond the circumstances that enable this to happen, the importance and need for instructors to take charge of producing such materials for their own use are simply that "teacher-made" resources are more impactful than externally produced material [33]. This is because the instructor is in control of ensuring the material is pedagogically appropriate, content accurate, fits within the desired learning progression, uses approaches and language suited to the learners, and is ultimately designed to bring about the desired learning outcomes. Nonetheless, this relatively nascent and as yet to be familiar technology in the classroom may present challenges to teachers in various ways, especially in terms of how best to employ it in pedagogically appropriate ways.

2 Out-of-Classroom Learning and Laboratory Practical Work

Organized OOC learning activities and science laboratory-based practical activities are integral components of science education curricula at all levels and are widely considered to be important pedagogical approaches to the teaching of science [11, 12, 21, 26]. Out-of-classroom learning and science laboratory practical support and extend traditional classroom academic learning by providing embodied hands-on *experiential learning* by *doing* and/or by *being there* [15, 18].

2.1 Out-of-Classroom Learning

OOC learning encompasses a broad and somewhat loosely defined collection of learning approaches that essentially involve learning outside of formal lessons conducted within a traditional classroom setting. OOC learning could range from structured lessons organized by an instructor as part of a formal curriculum, to informal everyday experiences encountered by the learner on their own time. The former might be exemplified by an organized field trip to a museum and the latter by a child observing that ice melts faster on a hot surface. In this chapter, the use of the term OOC learning focuses is on the learning of science in generally more structured approaches that either complement or are part of a formal curriculum. In other words, learning activities such as a field trip or visit to a science place-of-interest (POI), and online, multimedia, or video learning material that students might engage with outside of school hours. Examples of science POIs are museums, zoos, nature reserves, farms, factories, or sites with phenomena or artefacts of scientific interest, such as geysers or fossils.

OOC learning in science links formal classroom learning with the informal, everyday learning of the natural world around us [30]. It can provide the real-world context or examples covered in lessons and allow learners to explore and learn with greater *agency*. For example, individual learners may choose to spend more time engaging with specific exhibits that fascinate them at a museum. It is also well known to educators that field trips are exciting, especially for younger learners who will look forward to such activities, and that this strong sense of *engagement* and *motivation* is important aspects of what makes this form of learning productive and effective [12]. Such field trips are important contributors to *place-based education* [29], wherein the *sense of place* is a key pedagogical feature that "roots" the learner in the local context or community so that they may learn what is desired in or about that place.

2.2 Laboratory Practical Activities

Science practical work, where learners engage in activities where they interact with, manipulate, and make observations of physical objects, events and/or phenomena, has been widely used for a long time and is seen as an essential feature of science education [1]. Like OOC activities, practical work can describe a broad and diverse set of learning activities, some of which overlap with OOC learning in the form of field work. Here, we have chosen to focus on laboratory-based practical work undertaken as part of structured, formal lessons in schools, or tertiary institutions of learning, that is, the type of learning activities normally associated with the science laboratory in school settings. These are often of a *hands-on* nature where the learner must individually perform procedural tasks in order to engage in some aspect of scientific practice, such as investigating, observing, inferring, or analysing. Often inaccurately referred to as "conducting science experiments", such science practical work is typically intended to demonstrate to the learner some concept or phenomenon, and through the *embodied* actions of carrying out the procedure, learn about the *ways of doing* science. More recently, there has also been an increasing focus on getting students to learn the nature of science, its epistemological underpinnings, and the cooperative endeavour that is science research, in short, getting learners to learn science by engaging in the activities of scientists.

A key challenge in the teaching of science through practical work is communicating the necessary *procedural knowledge* for the "doing" of science, as opposed to the declarative knowledge of theoretical concepts and ideas. The former is often difficult to verbalize or formally dictate in written form and usually requires instructor demonstration, expert guidance or coaching, and the learner's practice in order to acquire that *skill*. Obviously, learning to play soccer if you were only given a book about it would be ludicrous, but learning just by watching video-recorded matches would only be a little better. As a performative set of skills to be learnt, the learner would be expected to be present with the coach, watching demonstrations at close quarters from various vantage points around but still within the same *space* as the coach, then engaging in practice with expert guidance, so too, in many ways similar, does the learner learn skills in the science laboratory.

2.3 *Common Problems with OOC and Practical Work*

In the routine execution of OOC, place-based learning, and science practical work, there are common difficulties faced by instructors, due at least in part to the need for learners to have particular vantage points with respect to the instructor, the location, and/or whatever they are required to engage with.

During science lessons involving field trips, the number of learners, location constraints, environmental, or situational impediments, such as distracting ambient noise, or the need to keep relatively quiet in certain venues, often contributes to difficulties for all learners from a large group to be able to simultaneously listen to their instructor. For example, in a forested nature reserve, visitors may be required to remain along a narrow trail. If the instructor pointed to some specimen and spoke about it, only those learners in the immediate vicinity would be able to see and hear the instructor clearly. Shouting or using voice amplification would be counterproductive as it would disturb the fauna and opportunities to observe specimens would be lost. For novice learners, making scientific observations in the real-world environment for specimens, phenomena or events can be overwhelming as it can be difficult to know "where to start" and "where to look". In laboratory-based practical lessons, similar issues may be present, especially when the instructor carries out demonstrations. Typical approaches to ameliorate such challenges include the use of tools such as a loudhailer, pre-activity briefings with photo or video footage, and/or with logistical arrangements such as rotating students through station-based instruction.

However, such measures may not always be appropriate nor adequately address these problems. As experiential approaches to learning, aspects of field trips and laboratory practical work can be difficult to reproduce or represent in still images or regular video footage. The sense of "being there" and being able to acutely observe the entire surroundings and/or events is typically lacking. But it is exactly this need for an embodied presence (or sense of it) that is typically used as justification for the importance of field trips and practical work in science education [17, 21, 33]. Supporting such learning activities, or indeed potentially replacing some portion of them with alternative approaches, thus, requires technological solutions that provide such perceptions of *presence*.

3 VR360 in Science Education

3.1 *Applications of VR360 in Science Learning*

As science educators and science education researchers, we originally envisaged the use of VR360 as primarily an additional, adjunct tool that could be employed *in support* of conventional place-based learning modes, and in particular, those of OOC learning and in science laboratory practical work. It is a common practice for instructors to demonstrate science laboratory practical work prior to learners themselves engaging in the tasks. For field trips, it is also known that pre-visit preparation and familiarization, and post-visit activities positively enhanced learning outcomes [19].

3.1.1 How Can VR360 Be Used?

In that conception, the key use-case for VR360 media is as pre-activity briefing or familiarization material for learners prior to embarking on OOC and laboratory-based practical activities. It has been found that VR360 media used in this manner is typically useful for this purpose although not without certain issues [4, 23, 27]. Essentially, VR360 images and video footage are made available to learners either as individual playable media clips or as part of ILE packages created using VR360 authoring platforms. For OOC field trips, learners can familiarize themselves with the environment of the venue, the appropriate routes and precautions to take, the key observations to make at various points, and any other instructions that the instructor may typically give before or during such visits (see Fig. 1). An example clip created by the authors can be viewed at https://youtu.be/allY0IzGI8c [32]. A version with informational hotspots created with interactive tour software is accessible on request. For laboratory practical work, key safety precautions can be emphasized, procedural methods can be demonstrated, and information such as names of various apparatus, reagents, and instrumentation can be shown (see Fig. 2).

The secondary use-cases for VR360 media may thus be (a) as post-activity debriefing material, where the VR360 media is played again to aid in recall of specific locations or events to reinforce learning, assessment or other aspects of

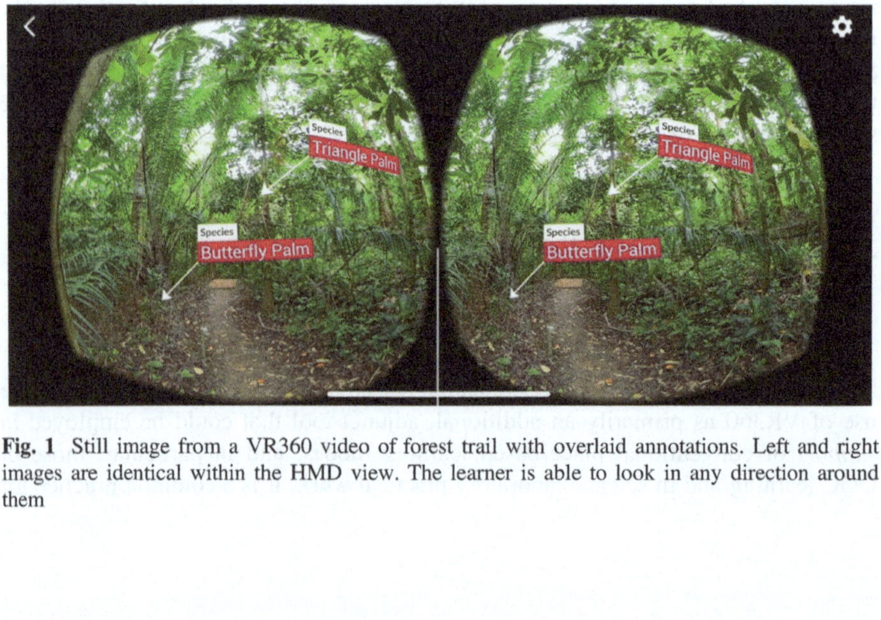

Fig. 1 Still image from a VR360 video of forest trail with overlaid annotations. Left and right images are identical within the HMD view. The learner is able to look in any direction around them

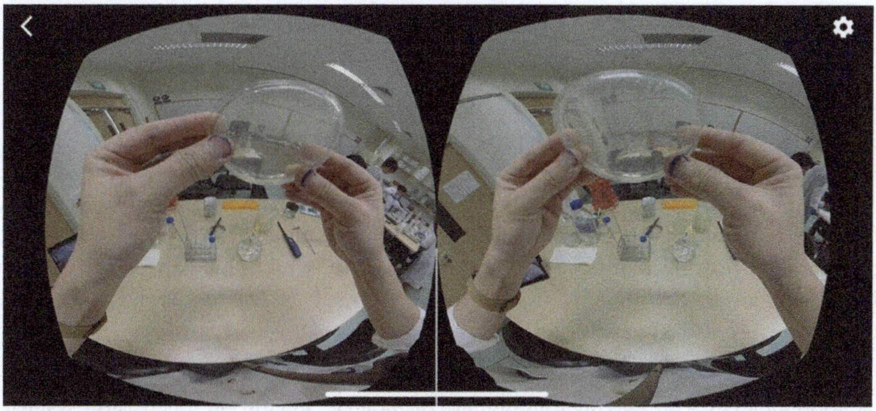

Fig. 2 Still image from a stereoscopic VR180 FPV video of a microbiology task. Note the difference in perspective between the left and right images. The learner can look around within a 180° hemisphere

lesson closure; (b) as post-activity debriefing material using VR360 media live recorded during the particular instance of that learning activity carried out in the real world; (c) as a *live stream* for real-time viewing of instructor demonstrations and exploration of field locations in situations where students may not be able to congregate in-person due to lack of space, accessibility, safety, or other considerations; and, essentially what was originally thought of as a last choice option, (d) as a recorded substitute for the real-world activity, for reasons such as inclement weather forcing cancellation of a field trip, where the activity is hazardous or unfeasible to undertake, for viewing by learners absent from the activity, or where the phenomena, specimen, or event is rare and unlikely to be encountered.

Prior to the COVID-19 pandemic, the last of these use-cases (where the ILE entirely replaces the real-world learning activity) was to the majority of science educators, unlikely to be seen as a desirable substitute for learning in OOC and science practical work when used in isolation. The embodied and hands-on interaction afforded by these modes of science instruction does not seem replaceable through other means. However, the massive and currently ongoing shift in circumstances has thrust virtual ILEs into the pedagogical limelight, and suddenly, the last option becomes the default option for instruction when disease control and safe/physical distancing requirements enforce a shift to remote learning modes. This will undoubtedly spur development and adoption of ILE in its various forms for education, but fundamental pedagogical issues and challenges remain with what is still a nascent tool for teaching and learning. As a counterpoint, some recent work has suggested that the palette of technological features available to the instructor, for example in the combined use of VR360 media with mixed spatial orientations (FPV and pseudo-aerial views), can actually enhance the embodied experience on virtual field trips [17], thus suggesting that virtual replacements for real-world learning experiences *can* be made effective through appropriate design and implementation considerations that leverage on the new affordances enabled by the technologies themselves.

3.1.2 How Can VR360 Be Deployed for Teaching and Learning?

As the focus of this chapter is on the development of VR360 learning resources by teachers and instructors themselves, a search for suitable software tools, media creation and hosting platforms, and other means by which this can be achieved at low to moderate cost and by the educators themselves was undertaken. The key criteria for inclusion were as follows:

(a) Access to software and/or service by the media creators (instructors) online or in downloadable on personal computers or devices, ideally on a self-service basis.
(b) VR360 resources that once created can be shared with an audience of learners, either online, downloadable, or installed in a suitable viewing device.

Services or vendors without specific VR360 support, those only providing turnkey or bespoke custom-coded solutions, or that did not showcase or demonstrate an existing ready-to-use platform, were excluded. The search was conducted in late 2019 and updated in June 2020. Services and software matching these criteria are listed in Table 1.

At the time of this writing, it can be seen that there are four general methods to deploy low-cost VR360 learning resources available to an instructor:

(a) *Offline media*: Individual VR360 images and/or video are pre-loaded on suitable personal computing devices or HMDs and viewed by the learners. Any annotation or editing of the media must be done manually by the instructor using common image or video editing software. No interactivity other than playback control.
(b) *Media-sharing platform*: Uploaded VR360 images and/or video is viewable online by learners only as individual media clips. Minimal, if any, means to add hyperlinks or hypermedia. Any annotation or editing of VR360 media must be done prior to upload as above. Examples: *YouTube*, *Vimeo* and *Flickr VR*.
(c) *Virtual tour platform*: Uploaded VR360 images and/or videos may be annotated, linked together, edited, and have hyperlinks/hypermedia embedded to various degrees within the platform. Each image or clip serves as a *scene* within a tour. Learners access via a virtual tour interface with the ability to move from scene to scene and interact with hypermedia, if present. A few enterprise-level platforms also incorporate hosting of AR/VR/MR. Examples: *Google Tour Creator*, *360°*, and *EON Reality AVR platform*.
(d) *VR360 authoring tools*: Software programs that create HTML5-based packages from VR360 images and/or video, similar to the online virtual tour platforms above. The HTML5 packages may be viewed offline via a *localhost* web server or uploaded and hosted on a suitable personal or institutional web server. Examples: *3DVista Virtual Tour*, *Pano2VR*.

Table 1 Platforms and software for publishing VR360 media

			Pricing (USD)		VR360 media		Features	
Platforms	Focus	Access[1]	Model	Annual fee	Image	Video	HMD Support	Interac-tivity
Online Service Providers								
YouTube	Media sharing	SS	Free	–	✓	✓	✓	+
Vimeo	Media sharing	SS	Freemium	From 84	✓	✓	✓	–
Flickr VR	Media sharing	SS	Freemium	From 72	✓	–	✓	–
Google Tour Creator	Virtual tours	SS	Free	–	✓	–	✓	+ +
Roundme	Virtual tours	SS	Freemium	99	✓	–	✓	+ +
GoGira360	Virtual tours	SS	Freemium	From 180	✓	–	–	+ +
Kuula	Virtual tours	SS	Freemium	From 144	✓	–	–	+ +
360°	Mixed	SS	Freemium	From 360	✓	✓	✓	+ +
Thinglink	Mixed	SS	Free trial	From 1200	✓	✓	✓	+ +
EON Reality AVR	Mixed	E	Free trial	On request	✓	✓	✓	+ + +
Omnivirt	Advertising	E	Free trial	On request	✓	✓	✓	+ + +
Viar360	Training	E	Free trial	From 1188	✓	✓	✓	+ + +
Desktop Authoring Software								
3DVista Virtual Tour	Virtual tours	SS	Free trial	570 (one-time)	✓	✓	✓	+ +
Lapentor	Virtual tours	SS	Freemium	From 10/project	✓	–	✓	+ +
Marzipano	Virtual tours	SS	Free	–	✓	–	–	+ +
Pano2VR	Virtual tours	SS	Freemium	400 (one-time)	✓	✓	✓	+ + +

[1]Access: *SS*—Self-Service, *E*—Enterprise. Information correct as of 15 June 2020

These methods are generally suited for use with personal computing devices with or without a HMD, mobile devices, and stand-alone HMDs with playback using a suitable media player or browser. Indeed, it would be expected that class size groups of learners can simultaneously view such VR360 resources using mobile phones, perhaps mounted in HMD adapters such as *Google Cardboard* frames.

3.2 Affordances of VR360 for Science Learning

Technology and media have long played important roles in supporting teaching and learning. The use of *multimedia* in science education as a mix of text, speech, sound, graphics, animation, and video is commonplace and useful to help learners understand concepts, relationships, and often to let them vicariously observe or experience various phenomena.

In common with such conventional sources of instructional media, VR360 media may afford both the learner and the instructor:

(a) Creation of reusable learning resources with known, reproducible content that can be produced, curated, and presented in pedagogically appropriate ways to the learner.
(b) The ability to incorporate textual, audible, graphical, or other forms of media within the environment of the immersive media, either as overlays, hotspots, popups, or hyperlinks to additional information. The instructor or media producer could employ these in multiple ways, shown by default or interactively activated, used for informational, assessment or gamification purposes, and so on.
(c) The ability to view, observe, or experience events or phenomena that are rare and unusual, inaccessible or hazardous, involve very short or very long time-scales, or are otherwise difficult to do so in-person.
(d) The ability to engage in asynchronous, self-paced learning where the media could be viewed repeatedly, selectively, with variable playback speed, paused, enlarged, or otherwise manipulated for optimal viewing. Where learning of skills is the objective, the learner may also be given the opportunity to mentally or even physically rehearse required actions or procedural steps, while viewing the media.
(e) The ability to substitute, or function as a backup/fallback option, for actual learning activities that cannot take place.
(f) The ability to reach out to audiences beyond those who can be physically present. This does not only apply to a remote audience, but even in local situations where it is impractical to position all the learners physically within a limited space.

In contrast, VR360 immersive media affords certain headline features that are either absent or of lesser impact in conventional media-based resources:

(g) Chief among these is the user's experience of *presence* in VR360 media, created by placing the user within the enveloping view at the position of the recording camera, and enabling user autonomy to gaze in any direction [7]. This sense of presence is both an essential aspect of the user's *immersion* within the VR environment [13, 28], as well as being the affordance that allows VR360 to support or even substitute for OOC learning and certain aspects of laboratory practical work. Informal feedback received from users viewing a prototype VR360 video of a forest trail [32] suggests that the recorded ambient sound from birds and insects in combination with VR360 video images noticeably contributes to this immersion.

(h) In itself, the user's ability to look around and choose what to look at is a key affordance that supports learning by giving the learner the sense of *agency* and hence engagement and motivation as an active participant, as opposed to being a passive viewer. On the other hand, as compared to VR simulations, VR360 features much less user autonomy to control their location or vantage point, pace of the action and despite the potentially photorealistic visuals, lack the ability to interact with their surroundings or manipulate any "objects" within their field of view. This reduced *interactivity* compared to VR simulations is not necessarily a disadvantage from an instructional perspective. A teacher may not want students to continually interact with the surroundings and, thus, become distracted from the desired learning objectives. Limiting the learner's options to selected hotspots of curated information can focus attention and perhaps minimize time spent on exploration.

(i) For live recordings made for, or even during, OOC activities where the learners are on-site, one key affordance of VR360 over traditional video is the potential to capture phenomena and events that the learners may not have observed simply because they were not gazing in the appropriate direction at the time. Conventional video recordings are dependent upon where the camera was pointed at the time, resulting in a fixed viewport of a given time and space during playback. In VR360 media, images are captured in practically all directions, allowing the user to "point" or direct their viewport anywhere during playback. Watching a VR360 media clip, especially with a HMD, the viewer essentially experiences it in subtly but infinitely different ways each time, depending on where they gaze at any and every point during playback.

(j) With care to employ the recording of VR360 media with a first-person viewpoint, not only are the above benefits achieved, but where learning entails the learner's awareness of and/or orientation to spatial relationships, the affordance of FPV should facilitate learning. For example, experiencing the relative scale of the trees as the learner moves along a forest trail, or observing specific laboratory techniques that require some dexterity from the FPV as opposed to the mirror image perspective, the learner would normally have while observing an instructor. Where stereoscopic VR180 footage is employed, this affordance would be enhanced through a more naturalistic perception of depth.

3.3 *Combining VR360 with Other AVR Modalities for Science Learning*

ILEs can potentially comprise more than one form of AVR technology. Indeed, there are platforms that already allow some degree of cross-incorporation of one format within another, for example, where VR360 video is playable within a VR simulation, and where VR objects can inserted within VR360 media clips in hybrid interactive simulation platforms that blur the distinction between VR and VR360 [2]. Another interesting mashup may be the overlaying of VR360 imagery on the surroundings in a MR setting, to show, for example, what the environment might have looked like with flora and fauna from particular time periods.

One scenario in which VR360/VR180 video and interactive VR simulations can complement each other in a simple way with current technologies is to create both forms of media around the same ILE or context. In this way, each platform can be used in a role that engages their respective affordances and strengths, and when used in conjunction, presents to the learner a multi-modal approach to a topic. For example, the authors are developing an instructional module aimed at the undergraduate biology level on the topic of basic bacteriological culture laboratory techniques. VR180 stereoscopic FPV video is recorded of an expert performing the necessary procedures and steps that the learner is expected to be able eventually to do themselves (Fig. 2). This video is paired with an interactive VR simulation where the same laboratory environment, apparatus, and materials are presented, in which the learner must manipulate the virtual objects to replicate the laboratory techniques seen in the VR180 video. Used in this manner, the VR180 video serves as self-paced, *demonstration* material with a first-person viewpoint for learning purposes, while the interactive VR simulation provides the ability to apply and *practise* the techniques learnt. With the appropriate simulation logic incorporated, feedback can be given to the learner on the correctness of procedures done. Finally, the learner can undertake the hands-on procedures in the actual laboratory with greater confidence and familiarity.

VR360 video can similarly be paired with AR for OOC and place-based learning. As described earlier, VR360 video can be used for briefing and familiarization purposes prior to a field trip. But where it might not be practical to use VR360 *while* on the field trip, the information and annotations from the VR360 material could be used in the form of AR overlays viewed on devices such as personal mobile phones or tablets while walking the actual ground. The alignment of the same learning content presented across platforms, whether from VR180 to interactive VR, or from VR360 video to AR overlays, should allow the learner to easily link and recall the material.

4 Issues and Challenges with VR360

4.1 Pedagogical Issues

Just as various forms of remote and online modes of learning, so-called *e-pedagogies* have presented issues and challenges due to their inherent differences from traditional pedagogies developed for synchronous, interactive face-to-face instruction, so too does VR360 for ILEs. It might be argued that because VR360 seek to envelope the learner within a world that resembles or even replicates the real, that because such ILEs add immersive, embodied, and spatial cues to life-like audio-visual media, and hence VR360 ILEs are more accurate facsimiles to actual field trips and practical work, that therefore VR360 ILEs should have fewer pedagogical issues to grapple with than previous generations of electronic learning resources. However, certain issues inevitably persist, while additional ones arise.

4.1.1 Interactions

Video-based ILEs present a particular set of limitations to learner–instructor and learner–learner (peer) interactions. With standard video-based media, an instructor has the option to play the video on a screen for an entire class to view at the same time and be able to, for example, control appropriate pauses for interactions (responding to student queries and instructor's questioning). However, this is probably not a typical mode of delivery for immersive video. Instead, learners essentially must engage with VR360 on a self-paced, individual basis which thus tends complicate learning interactions with each other and with the instructor. Learner–instructor and learner–learner interactions are important aspects of learning and are persistent issues in non-traditional modes of learning, especially for asynchronous modes [22]. For example, it would be difficult for a learner to ask a question of the instructor, especially if using an HMD. Depending on the system used and whether the instructor and learner are co-located in time and space, it would not be easy for the instructor to know exactly what the student might be looking at when the question was raised, as the instructor may not have ready access to that particular learner's view within the ILE. As for learner–learner interactions, these cannot be synchronous nor interactive as they could be in their own individual virtual environments, and avatars of other learners do not appear within each other's video stream. The instructor must, thus, structure and manage the lesson so as to allow for queries, suitable questioning, and discussion at appropriate points after each segment of VR360 video.

4.1.2 Note-Taking

If an ILE presents information or content to the learner or requires the learner to *observe* some feature or phenomena in the scientific sense, how will they make appropriate and timely notes while in HMD-based ILEs? This otherwise trivial task, normally accomplished by writing or drawing on a sheet of paper, in a notebook or perhaps done with the aid of a device such as a phone or tablet, becomes somewhat difficult. There is typically no convenient way for an individual learner to use traditional means of note-taking without taking off the HMD and letting go of any controllers. If the learner is not stationary at a desk, there is also the additional problem of keeping the note-taking materials within reach and conveniently placed for writing or typing. The learner could of course make notes after an ILE session, but for longer sessions and where observations should be made "on the spot" for accuracy or clarity, waiting till after the session is not ideal. As video-based ILEs do not have a fully virtual and interactive environment around the learner, there would not be virtual representations of note-taking media, such as a virtual notepad or keyboard within the ILE. Even if there were provisions of such interfaces, current generation controllers do not make it easy nor efficient to make notes, record observations, or answers.

Recording of verbal note-taking might be one potential method that would be more amenable in such situations, but affordances and interfaces will have to be built into ILE platforms. There would also need to be ways in which to transcribe such audio notes into written form for subsequent reference. Existing voice-activated virtual digital assistants or agents such as Google Home, Amazon Alexa, or Apple Siri may possibly be useful conduits for this type of learning support, where learners could call on such agents to "remember" or "note" specified materials and even answer their queries via internet searches.

Interestingly, a simple, practical approach that addresses the problem of note-taking as well as that of peer interaction has been utilized by history teacher Kim Young, where her students work in pairs while sharing one VR360 viewer, taking turns to write down their partner's verbal observations made while viewing the VR360 video [35].

4.2 Authoring Challenges

In the current scan of the tools and platforms available to an educator eager to self-create VR360 learning resources (see Table 1), it was apparent that while there was ready access to cameras for capturing footage and a plethora of personal computing devices and HMDs for viewing purposes, the middleware or software tool chain connecting raw VR360 capture to finished media-rich and educative playback remains relatively underdeveloped. There are currently few offline photo and video editing software applications that can natively work with VR360 media in an easy-to-use manner. For instance, the trivial task of overlaying a text label to a flat image becomes a complex task in VR360 because the label has to conform to the spherical geometry of the 360° image. We also see few online platforms and services that are deeply oriented to the needs of education. The present feature sets have been developed in response to the needs of commercial interests, such as in virtual tours for the property market, and even where the focus is on education and training purposes, the toolset is still fairly rudimentary, making desirable pedagogical staples such as gamification, assessment and feedback, and other aforementioned issues relatively difficult to implement.

Other challenges are common or inherent to the uptake of nascent technologies abound but are beyond the scope of this chapter. Suffice to hope that VR360 may one day be as ubiquitous, as easy to create, and as easily deployed in the classroom as slideshow presentations are today.

5 Conclusions

VR360 technology holds exceptional potential to support the improvement, and possibly even the transformation, of the learning of science, particularly in the areas of OOC learning and practical work. The technology is already accessible today to educators at all levels with the easy and affordable availability of the necessary hardware, software tools, and an already robust technological ecosystem for the production and deployment of VR360-based media and ILEs. Compared with simulation-based VR technologies, VR360 shares many features, such as providing an immersive experience and allowing for repeatable but also variable views of the content, but at a substantially lower development cost, with much quicker and easier production, which in turn allows for timely creation of such VR resources. In circumstances such as the dramatic and sudden shift towards remote learning due to pandemic control measures, this low entry threshold enables individual educators to rapidly create VR content to substitute for particular OOC or other instructional needs.

These advantages of VR360 do come somewhat balanced against reduced interactivity with the VR environment and/or other users when compared to VR simulations. The user is typically limited to media playback controls and access to information via hotspots or simple triggers only. However, this in itself is not inherently a disadvantage but a difference in affordances that is important for an educator to note, to more effectively select among the various AVR modalities to suit the learning objective.

New technologies bring with them the potential for the development of hitherto unexplored pedagogical approaches, based on the novel affordances inherent within the technologies themselves. The particular combination of affordances from the use of real-life imagery, the enhancement of the learners' situated and embodied cognition through the sensation of presence, and the motivational and affective engagement provided by the sense of learning agency deliver a new pedagogical tool for experiential and/or place-based educational needs. VR360 already enables an instructor to overcome persistent issues encountered in OOC and science practical work learning activities now, and perhaps through novel application of and combinations with related AVR technologies, these could eventually further enhance learning outcomes.

Some specific pedagogical issues and challenges remain to be addressed and so too are other more broad-based ones in the greater e-pedagogies milieu, as the world grapples with dramatic shifts in emphasis on pedagogical modalities. But it is also this massive ongoing effort that is likely to rapidly and beneficially drive further innovation in VR360 technology for instructional purposes.

References

1. Abrahams, I., Millar, R.: Does practical work really work? A study of the effectiveness of practical work as a teaching and learning method in school science. Int. J. Sci. Educ. **30**(14), 1945–1969 (2008)
2. Adão, T., Pádu, L., Fonseca, M., Agrellos, L., Sousa, J.J., Magalhães, L., Peres, E.: A rapid prototyping tool to produce 360° video-based immersive experiences enhanced with virtual/multimedia elements. Procedia. Comput. Sci. **138**, 441–453 (2018)
3. Amadeo, R.: YouTube's "VR180" format cuts down on VR video's prohibitive requirements. Ars Technica (2017). https://arstechnica.com/gadgets/2017/06/youtubes-vr180-format-makes-vr-video-more-accessible/. Accessed 15 June 2020
4. Ardisara, A., Fung, F.M.: Integrating 360° videos in an undergraduate chemistry laboratory course. J. Chem. Educ. **95**(10), 1881–1884 (2018)
5. Argyriou, L., Economou, D., Bouki, V.: Design methodology for 360° immersive video applications: the case study of a cultural heritage virtual tour. Pers. Ubiquitous Comput. (2020). https://doi.org/10.1007/s00779-020-01373-8
6. Bassford, M., Crisp, A., O'Sullivan, A., Bacon, J., Fowler, M.: CrashEd–a live immersive, learning experience embedding STEM subjects in a realistic, interactive crime scene. Res. Learn. Technol. **24**(1) (2016). https://doi.org/10.3402/rlt.v24.30089
7. Bonner, E., Lege, R.: 360° Videos in the classroom: a how-to guide. Lang. Teach. **42**(5), 29–31 (2018)
8. Bonnington, C.: You can now watch and upload 360° videos on YouTube. Wired (2015). https://www.wired.com/2015/03/youtube-360-degree-video/. Accessed 15 June 2020
9. De Freitas, S., Rebolledo-Mendez, G., Liarokapis, F., Magoulas, G., Poulovassilis, A.: Learning as immersive experiences: using the four-dimensional framework for designing and evaluating immersive learning experiences in a virtual world. Br. J. Educ. Technol. **41**, 69–85 (2010)
10. DeAngelis, T.: A second life for practice?. Monit. Psychol. **43**(3) (2012). https://www.apa.org/monitor/2012/03/avatars. Accessed 15 June 2020
11. Hofstein, A., Kind, P.M.: Learning in and from science laboratories. In: Fraser, B.J., Tobin, K., McRobbie, C. (eds.) Second International Handbook of Science Education, Springer International Handbooks of Education 24. Springer, Dordrecht (2012)
12. Hofstein, A., Rosenfeld, S.: Bridging the gap between formal and informal science learning. Stud. Sci. Educ. **28**, 87–112 (1996)
13. Ijsselsteijn, W.A., Riva, G.: Being there: the experience of presence in mediated environments. In: Riva, G., Davide, F., Ijsselsteijn, W.A. (eds.) Being There: Concepts. Effects and Measurement of User Presence in Synthetic Environments. Ios, Amsterdam (2003)
14. Info-communications Media Development Authority, Singapore (n.d.): Immersive media–technology and R&D roadmaps. https://www.imda.gov.sg/-/media/imda/files/industry-development/infrastructure/technology/technology-roadmap/immersive-media.pdf?la=en. Accessed 8 Feb 2019

15. Jose, S., Patrick, P.G., Moseley, C.: Experiential learning theory: the importance of outdoor classrooms in environmental education. Int. J. Sci. Educ. Part B: Commun. Public Engagem. **7**(3), 269–284 (2017)
16. Kavanagh, S., Luxton-Reilly, A., Wuensche, B., Plimmer, B.: A systematic review of virtual reality in education. Themes Sci. Technol. Educ. **10**(2), 85–119 (2017)
17. Klippel, A., Zhao, J., Oprean, D., Wallgrün, J.O., Stubbs, C., La Femina, P., Jackson, K.L.: The value of being there: toward a science of immersive virtual field trips. Virtual Real (2019). https://doi.org/10.1007/s10055-019-00418-5
18. Kolb, D.A.: Experiential Learning: Experience as the Source of Learning and Development. Prentice-Hall, New Jersey (1984)
19. Lee, H., Stern, M.J., Powell, R.B.: Do pre-visit preparation and post-visit activities improve student outcomes on field trips? Environ. Educ. Res. (2020). https://doi.org/10.1080/13504622.2020.1765991
20. Leong, C.: Singapore's digital economy gunning for four key areas (2017). https://www.imda.gov.sg/infocomm-and-media-news/buzz-central/2017/6/singapores-digital-economy-gunning-for-four-key-areas. Accessed 8 Feb 2019
21. Millar, R.: The Role of Practical Work in the Teaching and Learning of Science. High School Science Laboratories: Role and Vision. National Academy of Sciences, Washington, DC (2004)
22. Moore, M.G.: Three types of interaction. Am. J. Distance Educ. **3**(2), 1–7 (2009)
23. O'Sullivan, B., Alam, F., Matava, C.: Creating low-cost 360° virtual reality videos for hospitals: a technical paper on the dos and dont's. J. Med. Internet Res. **20**(7), e239 (2018). https://doi.org/10.2196/jmir.9596
24. Pagano, K.O.: Immersive Learning. American Society for Training and Development, Alexandria, VA (2013)
25. Radiantia, J., Majchrzak, T.A., Fromm, J., Wohlgenannt, I.: A systematic review of immersive virtual reality applications for higher education: design elements, lessons learned, and research agenda. Comput. Educ. **147**, 103778 (2020)
26. Rennie, L.J.: Learning science outside of school. In: Lederman, N.G., Abell, S.K. (eds.) Handbook of Research on Science Education. Routledge, New York, NY (2014)
27. Rupp, M.A., Odette, K.L., Kozachuk, J., Michaelis, J.R., Smither, J.A., McConnell, D.S.: Investigating learning outcomes and subjective experiences in 360° videos. Comput. Educ. **128**, 256–268 (2019)
28. Slater, M., Wilbur, S.: A framework for immersive virtual environments (FIVE): speculations on the role of presence in virtual environments. Presence: Teleoperators Virtual Environ. **6** (6):603–616 (1997)
29. Sobel, D.: Place-based education: connecting classrooms & communities. Orion Society (2004)
30. Stocklmayer, S.M., Rennie, L.J., Gilbert, J.K.: The roles of the formal and informal sectors in the provision of effective science education. Stud. Sci. Educ. **46**(1), 1–44 (2010). https://doi.org/10.1080/03057260903562284
31. Sundar, S.S., Kang, J., Oprean, D.: Being there in the midst of the story: how immersive journalism affects our perceptions and cognitions. Cyberpsychology Behav. Soc. Netw. **20** (11), 672–682 (2017)
32. Tan, T.T.M., Seman, D., Tan, A.L.: Palms Along Wallace Trail (2020). https://youtu.be/allY0IzGI8c
33. Tuthill, G., Klemm, E.B.: Virtual field trips: alternatives to actual field trips. Int. J. Instr. Media **29**(4), 453–468 (2002)
34. Vettehen, P.H., Wiltink, D., Huiskamp, M., Schapp, G., Ketelaar, P.: Taking the full view: how viewers respond to 360° video news. Comput. Hum. Behav. **91**, 24–32 (2019)
35. Young, K.: Strategy share: using 360° technology to engage student explorers (2018). https://blog.education.nationalgeographic.org/2018/11/20/strategy-share-using-360-degree-technology-to-engage-student-explorers/. Accessed 15 June 2020

Efficient Facial Reconstruction and Real-time Expression for VR Interaction Using RGB-D Videos

Hua Ren and Xinyu Zhang

Abstract We present an efficient face reconstruction and real-time facial expression for VR interaction driven by RGB-D videos. A RGB-D camera is first used to capture depth images, and a coarse face model is then rapidly reconstructed. The user's personalized avatar is generated using pre-defined face model template and shape morphing techniques. We track the user's head motion and face expression using a RGB camera. A set of facial features are located and labelled on the colour images. Corresponding facial features are automatically labelled on the reconstructed face model. The user's virtual avatar is driven by the set of facial features using Laplacian deformation. We demonstrate that our algorithm is able to rapidly create a personalized face model using depth images and achieve realtime facial expression for VR interaction using live RGB videos. Our algorithm can be used in online learning environments that allow learners to interact with simulated and controlled virtual agents.

Keywords Virtual Reality · Online education · Facial recontruction · RGB-D videos

1 Introduction

Tracking facial expression is the key component in many applications like filming, virtual reality, human–machine interaction, gaming. Especially, for some highly interactive scenarios like virtual conference, virtual education and virtual social communication, real-time facial expressions help convey realistic characteristics of a physical person. Professional face tracking devices are able to obtain facial

H. Ren · X. Zhang (✉)
Shanghai Key Laboratory of Trustworthy Computing, School of Software Engineering, East China Normal University, Shanghai, China
e-mail: xyzhang@sei.ecnu.edu.cn

X. Zhang
School of Software Engineering, East China Normal University, Shanghai, China

Y. Cai et al. (eds.), *Virtual and Augmented Reality, Simulation and Serious Games for Education*, Gaming Media and Social Effects,
https://doi.org/10.1007/978-981-16-1361-6_14

expression in high precision. Then, an avatar is animated using expression transfer by aligning captured data to pre-defined face template.

There are two major approaches in tracking facial deformation: marker based and markerless. Marker-based tracking approaches require manually sticking markers on a user's face. Then, many work studied tracking markers' movements with high-resolution cameras [11, 22, 28]. For example, VICON optical motion capture system is a widely used device. However, such motion capture devices are too expensive to be suitable for massive users. A few studies used low-cost cameras to track markers [5, 19]. In addition, many markerless algorithms were developed [1, 8, 32]. They detected visual facial features and used them to guide facial deformation.

With the increasing popularity of inexpensive RGB-D cameras like Microsoft Kinect, Intel RealSense (e.g. some desktop/laptop computers like Dell, Lenovo and Asus are equipped with RealSense RGB-D cameras.), Asus Xtion Pro Live, etc., it becomes possible for massive users to generate their personalized avatars. Then, through their facial expression, users can be immersed in a virtual environment and create real-time social VR interaction and communication.

RGB-D cameras were used to record pre-defined example expressions [31] and later used for reconstruction by matching the user-specific expression models to the acquired RGB images and depth maps. However, it is time consuming to build such user-specific expression models. Consumer-level RGB-D devices were used in [7] to capture facial dynamics to animate virtual avatars. Many work [7, 14, 17, 31] employed 3D blendshape model to offer a compact representation for real-time tracking. It requires professional animators to create template blendshape models. Moreover, dynamic expression template may not be suitable for all age groups. A detailed 3D head model can be reconstructed using RGB-D camera [10, 26], and then, facial motion is tracked in real time using a RGB-D camera. This does not require complex pre-processing and modelling.

In this chapter, we present a real-time algorithm for facial animation driven by RGB-D videos. With an inexpensive depth camera, we are able to rapidly reconstruct a personalized virtual avatar using a pre-defined face model template and shape morphing techniques from captured depth images. Then, using a RGB camera, a set of facial features are located and labelled on the RGB images. Corresponding facial features are automatically labelled on the reconstructed face model assisted by a pre-defined face model template. A virtual avatar is driven by facial features and Laplacian deformation. In online learning environments, our algorithm allows learners to interact with simulated and controlled virtual agents.

We organize the rest of this chapter as follows. In Sect. 2, we briefly review prior work on virtual avatar reconstruction and face animation. Section 3 is the overview of our algorithm. In Sect. 4, we describe the algorithm of creating a personalized avatar. We explain the procedure of detecting a set of facial features and how corresponding facial features are automatically labelled on the reconstructed face model. In Sect. 5, we present the detailed facial expression transferring algorithm. We show our implementation in Sect. 6. Finally, we conclude this chapter and propose potential future work in Sect. 7.

2 Related Work

There are extensive studies on facial animation. Here, we briefly introduce some related work.

Face reconstruction can be used to rapidly generate a person's virtual avatar using colour images or depth images. Liu et al. [20] reconstructed a human face model from videos with user's manual interaction. A user manually clicks on the images and tells where the eyes, mouth and nose are. Cao et al. [9] created an image-based avatar from a set of sparse images captured from a Web camera. Blanz et al. [6] derived a morphable model from a 3D face model database. Creating a 3D avatar from a handheld video input was introduced in Ichim et al. [18]. The advent of consumer depth sensors such as the Microsoft Kinect led to some new approaches to 3D reconstruction. Chen et al. [12] incrementally deformed a 3D template mesh to match depth images and feature points detected from RGB-D images to create a user's avatar. Wang et al. [29] constructed a 3D model using either Possion surface reconstruction or KinectFusion. Zollhöfer et al. [34] presented a combined hardware and software solution for markless reconstruction. Some work used both colour images and depth images [33]. Sturm et al. [24] used KinectFusion to reconstruct 3D head model when a user sits on a swivel chair and turn 360° in front of the RGB-D camera.

Facial animation includes three types of techniques: direct parameterization, blendshapes interpolation and muscle actuation. A parametric facial animation defines a set of expression parameters for different face parts. Each expression parameter affects a set of vertices of the face model [11, 16]. Blendshape interpolation can be used to subtly adjust a facial model when an avatar's facial expression is compared with the data in a face model database [13, 31]. The movement of face muscles underneath skin has a direct influence on facial expression. Some work uses muscle actuation as the parameter to control facial animation [22, 30]. We refer readers to the survey in Deng and Noh [15] for more work.

3 Overview

Our real-time algorithm is designed for VR facial interaction and driven by live RGB-D videos. Our algorithm has two major components: personalized avatar reconstruction (see Fig. 1) and real-time facial interaction using live RGB videos (see Fig. 2).

Personalized Avatar Reconstruction (see Sect. 4). We first obtain a set of depth images captured from different views. Then, we convert each depth image to a point cloud. We align and merge all these point clouds into a single 3D facial point cloud model. At the same time, we obtain a corresponding RGB image and detect the

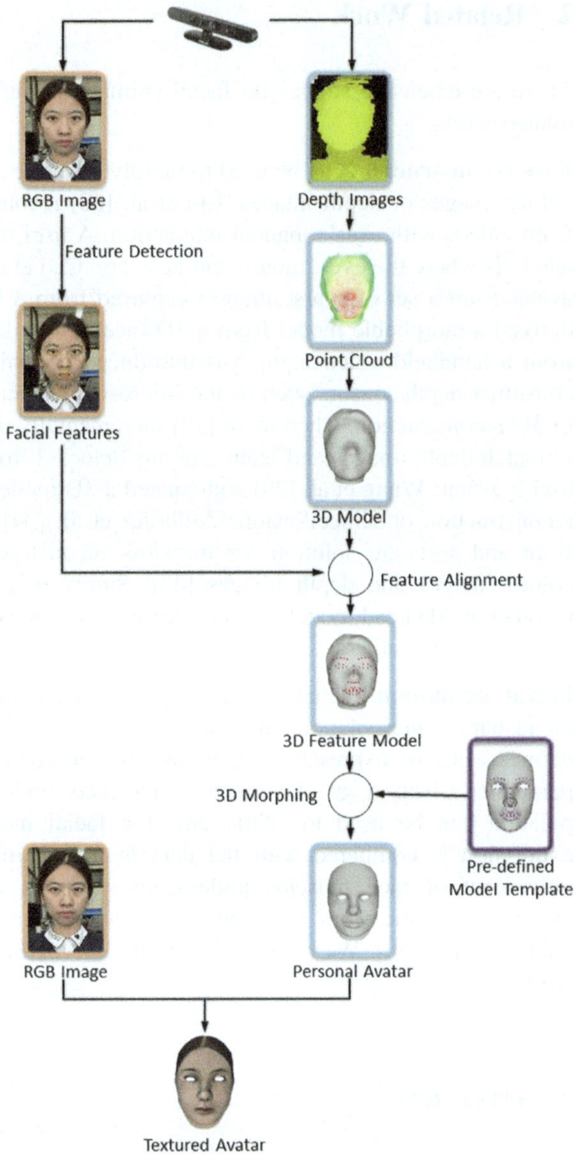

Fig. 1 Our real-time algorithm for VR facial interaction driven by live RGB-D videos. A personalized avatar is first reconstructed using both depth images and RGB images

facial features. Using these facial features, we identify the corresponding features in the 3D facial point cloud model. We deform a pre-defined head model template so that it gradually aligns 3D facial point cloud model. Finally, we apply RGB images to the resulting 3D facial model and generate a personalized avatar with texture. Refer to Fig. 1.

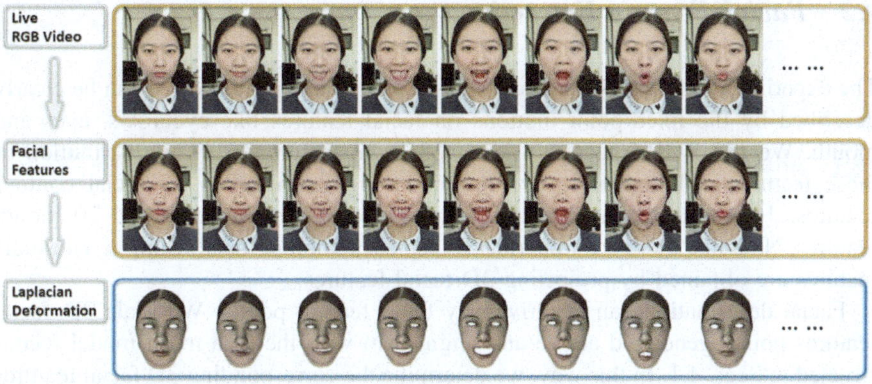

Fig. 2 Virtual avatar is driven by a set of facial features and Laplacian deformation in realtime

Realtime Facial Interaction (see Sect. 5). Using a live RGB video, we detect a set of facial features and rigid head motion. Then, we generate the virtual avatar's facial expression using these sparse features and Laplacian deformation. Refer to Fig. 2.

4 Personalized Avatar

In this section, we elaborate the details of generating a personalized avatar with texture using both depth and RGB images.

4.1 Avatar Face Acquisition

As shown in Fig. 1, for a fixed depth camera, a user is asked to rotate his/her head and a few depth images from different views are captured. We convert these depth images into point clouds [27]. The initial point clouds often contain noises so that we first remove noises using Gaussian filter. We use a voxel grid filter to subsample the point clouds. A 3D voxel grid consists of a set of cubic cells, and each cubic cell has a pre-specified size. We take the average of all the points in the cubic cell as the new point. After filtering all cubic cells in the 3D voxel grid, a new point cloud is generated. For a set of point clouds, we use iteration closest point (ICP) algorithm to align and merge them together into a single-point cloud model. The top row in Fig. 1 shows a single-point cloud model and its tessellation using ball pivoting algorithm [4].

4.2 Facial Feature Extraction

The experiment in [3] showed that the visual characteristics of a face can be clearly described by the fixed-point motions on facial features like eyebrows, eyes and mouth. We use RGB images to detect 68 facial features [2]. The distribution of these features and landmarks are shown in Fig. 3, including 17 facial contour features, 10 eyebrow features, 12 eyes features, 9 nose features and 20 mouth features. Note that there is an internal 3D representation of facial features. These 2D features are obtained by projecting 3D facial features.

Facial deformation can be driven by these feature points. We scale 3D facial features points generated above and align them with the face mesh model reconstructed in Sect. 4.1. In this way, we determine the corresponding 3D facial features in a face mesh model.

Let F_i be a set of feature points and M_r be the model obtained in Sect. 4.1. We transform F_i and M_r into the same coordinate system, and we rotate and scale F_i to fit M_r. Then, we use *k*NN [21] to find the nearest points to F_i in M_r. Here, we use a k-d tree to organize points and quickly search a k-dimensional space for the nearest points. We denote the resulting model as M_0.

Facial feature points and the corresponding points on model are shown in Fig. 3. The automatic alignment and mapping of feature points are shown in Fig. 4.

4.3 Face Morphing

We pre-defined a standard face model template M_t, and its facial feature points are pre-specified. We build one-to-one mapping between M_0 and M_t so that we can perform morphing between these two models. The morphing process is to find a set of affine transformations so that the model template M_t results in a new model that matches M_0. The match proceeds by minimizing weighted sum of errors including

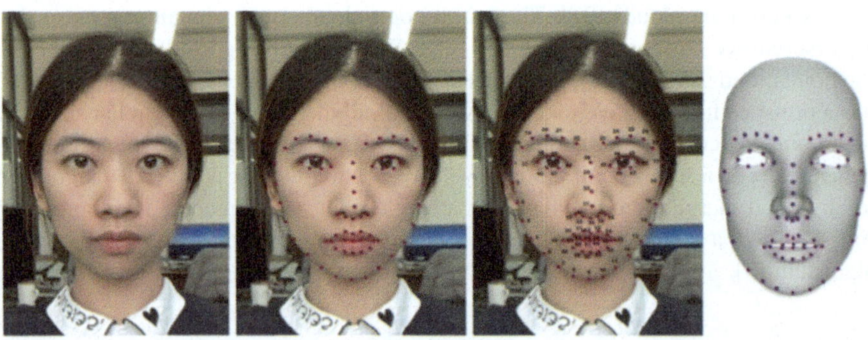

Fig. 3 Facial features

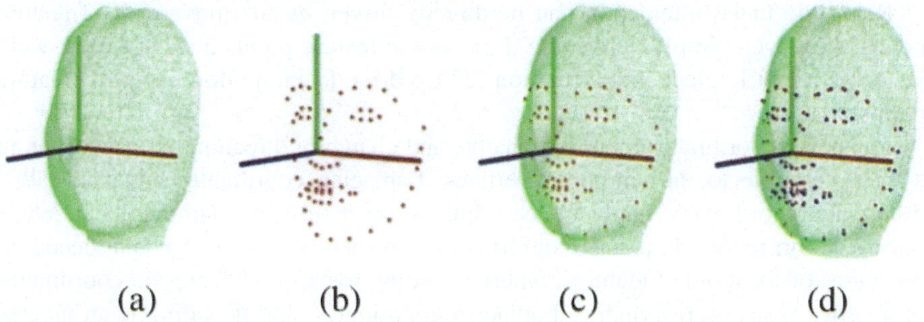

Fig. 4 Alignment and mapping. **a** A point cloud model; **b** feature points; **c** the alignment between the point cloud model and feature points; **d** the resulting point cloud derived by alignment and mapping

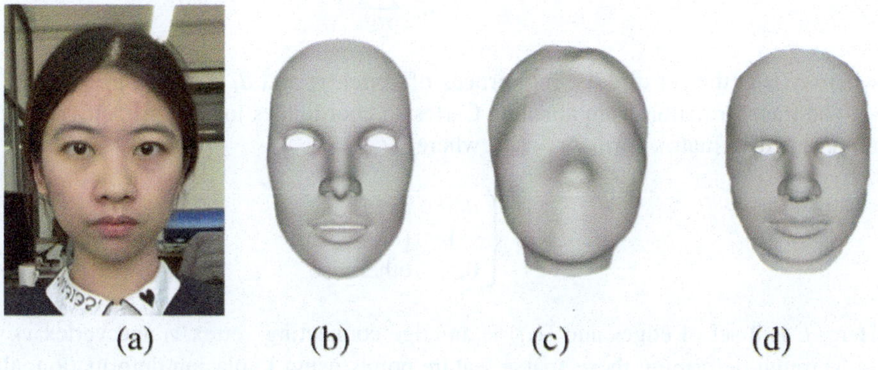

Fig. 5 Morphing between the model template M_t and the user's avatar. **a** A user's RGB photograph; **b** a pre-defined model template M_t; **c** a reconstructed coarse model M_0; **d** a personalized model after morphing

the data error and smoothing error subject to landmark error [25]. It is unnecessary that M_0 and M_t have the identical number of vertices or triangles or identical connectivity. Though the resulting model reconstructed in Sect. 4.1 has low resolution, we can still have high-resolution model using this morphing technique. This process is also shown in Fig. 5.

5 Real-time Facial Interaction

Once a personalized avatar is generated, real-time facial interaction can be achieved using live RGB videos and Laplacian deformation.

Real-time facial interaction can be directly driven by the movement of feature points. However, simply deforming these sparse feature points does not work well. Here, we adopt Laplacian deformation [23] to drive facial motion using 3D feature points.

Laplacian coordinates represent change and changing direction of each vertex in a mesh, relative to its neighbour vertices. Laplacian coordinates are also called differential coordinates. Laplacian coordinates can be used to maintain local details during deformation. Laplacian deformation consists of encoding and decoding processes of local detail features, where encoding transforms Cartesian coordinates of a vertex to its corresponding Laplacian coordinates, and decoding is an inverse process. Assume there are n vertices in a given model, where $V = (v_1, v_2, \ldots, v_n)$ represents the set of vertices and $v_i = (x_i, y_i, z_i)$ is Cartesian coordinates. The Laplacian coordinate δ of the vertex v_i is represented by the difference between v_i and the average of its neighbours.

$$\delta_i = v_i - \frac{1}{d_i} \sum_{j=N(i)} v_j$$

where $N(i)$ is the set of adjacent vertices of vertex v_i and d_i is the degree of v_i.

The transformation from absolute Cartesian coordinates to δ-coordinates can be represented in matrix form $\Delta = LV$, where

$$L_{ij} = \begin{cases} d_i, & i = j \\ -1, & (i, j) \in E \\ 0, & \text{otherwise} \end{cases}$$

Here, E is a set of edges and (i,j) is an edge connecting vertex v_i and vertex v_j.

Through deforming these sparse feature points using Laplacian deformation, all the rest vertices deform accordingly. Let

$$S = \{T, R, F_0, F_1, \ldots F_{67}\}$$

where $\{F_k\}$ are the set of 3D feature points ($k = 0, 1, \ldots, 67$) and $F_k = (x_k, y_k, z_k)$. T and R are the translation and rotation, respectively. After rotation R, the vertices of the model become $\left(V(t)^{'}\right) = R(t) \cdot V$. Then, Laplacian deformation will apply to $V(t)^{'}$. We define Laplacian coordinates $\Delta = L(V')$ at t.

We add 68 feature points into Laplacian matrix as anchors. Then, the size of the Laplacian matrix is $(n + 68) \times n$. According to formula $\Delta = L(V')$, we calculate the Laplacian coordinate Δ, where Δ is $(n + 68) \times 3$ matrix. If we let the last 68 rows of Δ be facial feature points (i.e. anchors), we have $\Delta^{n+k} = F_k, k = 0, 1, \ldots, 67$. To calculate the position of vertices V'' at t, we solve the following equation

$$(L^T L)V' = L^T \Delta$$

6 Implementation and Results

We implemented our system on a workstation with a quad-core Intel i7 3.4 GHz CPU, 16 GB memory and an NVIDIA GTX 1070 graphics card. All RGB images and depth images are captured with an off-the-shelf RGB-D camera at a resolution of 640 $\times$ 480. It takes about one minute for constructing a user's coarse model. To generate a personalized avatar, we choose a pre-defined model which contains 1219 vertices and 3728 triangles. It takes less than 1 min for reconstructing a personalized avatar from the pre-defined model template and the coarse model using

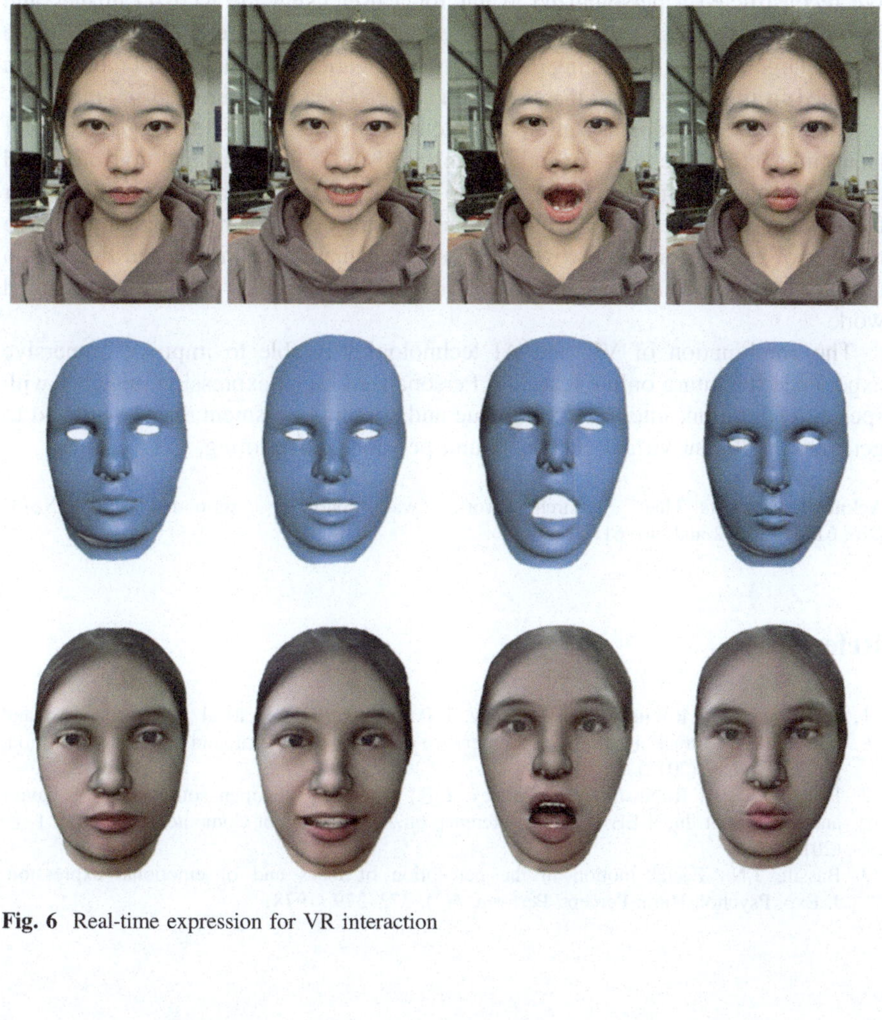

Fig. 6 Real-time expression for VR interaction

morphing technique. During the run time, it takes about 30 ms per frame to animate and render the avatar for an input video.

Figure 6 shows snapshots of the input RGB-D video, with different expressions: smiling, mouth opening and pouting. The second row shows the facial animation of corresponding avatar without texture. The third row shows the avatar rendered with user's facial texture.

7 Conclusions

Future online learning environments allow learners to interact with simulated and controlled virtual agents. Natural interaction, especially natural facial interaction, is able to help learners to immerse themselves in a virtual learning environment. We have presented a real-time algorithm for VR facial interaction driven by live videos. We reconstructed a personalized avatar for a user using the RGB-D images and morphing techniques. We do not require manual user interaction and nor do we require large amount of training data for avatar creation. With the facial motion data obtained from a live video, the personalized avatar can be animated in real time.

Compared to parameterization techniques, our method is more general. Our facial interaction does not depend on the facial mesh topology and does not need manually tune parameters. Compared to blend shape interpolation methods, our facial interaction uses Laplacian deformation and does not need creating a large number of expressions. Moreover, the muscle actuation methods need to create a muscle model for individual person, which often takes a lot of time and manual work.

The combination of VR and AI technologies is able to improve immersive experience for future online learning. Personalized facial expression, together with speech recognition, intellectual dialogue and speech assessment, can be utilized to generate intelligent virtual characters and personalized learning.

Acknowledgements This research work was partially supported by NSFC (No. 61631166002 and No. 61572196).

References

1. Baltrušaitis, T., Robinson, P., Morency, L.P.: 3D constrained local model for rigid and non-rigid facial tracking. In: IEEE Conference on Computer Vision and Pattern Recognition, pp. 2610–2617 (2012)
2. Baltrušaitis, T., Robinson, P., Morency, L.P.: Openface: an open source facial behavior analysis toolkit. In: IEEE Winter Conference on Applications of Computer Vision, pp. 1–10 (2016)
3. Bassili, J.N.: Facial motion in the perception of faces and of emotional expression. J. Exp. Psychol. Hum. Percept. Perform. **4**(3), 373–379 (1978)

4. Bernardini, F., Mittleman, J., Rushmeier, et al.: The ball-pivoting algorithm for surface reconstruction. IEEE Trans. Visual Comput. Graphics **5**(4), 349–359 (1999)
5. Bickel, B., Botsch, M., Angst, R., et al.: Multi-scale capture of facial geometry and motion. ACM Trans. Graph. **36**(3), 33–41 (2007)
6. Blanz, V., Vetter, T.: A morphable model for the synthesis of 3D faces. In: ACM SIGGRAPH, pp. 187–194 (1999)
7. Bouaziz, S., Wang, Y., Pauly, M.: Online modeling for realtime facial animation. ACM Trans. Graph. **32**(4), article 40 (2013)
8. Cao, C., Weng, Y., Lin, S., et al.: 3D shape regression for real-time facial animation. ACM Trans. Graph. **32**(4), article 41 (2013)
9. Cao, C., Wu, H., Weng, Y., et al.: Real-time facial animation with image-based dynamic avatars. ACM Trans. Graph. **35**(4), article 126 (2016)
10. Casas, D., Alexander, O., Feng, A. W., et al.: Rapid photorealistic blendshapes from commodity RGB-D sensors. In: ACM Symposium on Interactive 3D Graphics and Games, pp. 134–134 (2015)
11. Chai, J.X., Xiao, J., Hodgins, J.: Vision-based control of 3D facial animation. In: ACM SIGGRAPH/Eurographics Symposium on Computer Animation, pp. 193–206 (2003)
12. Chen, Y.L., Wu, H.T., Shi, F.H.: Accurate and robust 3D facial capture using a single RGBD camera. In: IEEE International Conference on Computer Vision, pp. 3615–3622 (2013)
13. Chuang, E., Bregler, C.: Performance driven facial animation using blendshape interpolation. Technical Report, Stanford University (2002)
14. Deng, Z., Chiang, P.Y., Fox, P., et al.: Animating blendshape faces by cross-mapping motion capture data. In: ACM Symposium on Interactive 3D Graphics and Games, pp. 43–48 (2006)
15. Deng, Z., Noh, J.: Computer facial animation: a survey. In: Data-Driven 3D Facial Animation, pp. 1–28. Springer, London (2008)
16. Goto, T., Escher, M., Zanardi, C., et al.: MPEG-4 based animation with face feature tracking. In: Computer Animation and Simulation, pp. 89–98. Springer, Wien (1999)
17. Huang, H., Chai, J.X., Tong, X., et al.: Leveraging motion capture and 3D scanning for high-fidelity facial performance acquisition. ACM Trans. Graph. **30**(4), article 74 (2011)
18. Ichim, A.E., Bouaziz, S., Pauly, M.: Dynamic 3D avatar creation from hand-held video input. ACM Trans. Graph. **34**(4), article 45 (2015)
19. Lin, I.C., Yeh, J.S., Ouhyoung, M.: Extracting 3D facial animation parameters from multiview video clips. IEEE Comput. Graph. Appl. **22**(6), 72–80 (2002)
20. Liu, Z.C., Zhang, Z.Y., Jacobs, C., et al.: Rapid modeling of animated faces from video. Comput. Animation Virtual Worlds **12**(4), 227–240 (2001)
21. Ramasubramanian, V., Paliwal, K.K.: Fast k-dimensional tree algorithms for nearest neighbor search with application to vector quantization encoding. IEEE Trans. Signal Process. **40**(3), 518–531 (1992)
22. Sifakis, E., Neverov, I., Fedkiw, R.: Automatic determination of facial muscle activations from sparse motion capture marker data. ACM Trans. Graph. **24**(4), 417–425 (2005)
23. Sorkine, O., Cohen-Or, D., Lipman, Y., et al.: Laplacian surface editing. In: Eurographics/ ACM SIGGRAPH Symposium on Geometry Processing, pp. 175–184 (2004)
24. Sturm, J., Bylow, E., Kahl, F., et al.: CopyMe3D: scanning and printing persons in 3D. In: German Conference on Pattern Recognition, pp. 405–414. Springer, Berlin (2013)
25. Sumner, R.W., Popović, J.: Deformation transfer for triangle meshes. ACM Trans. Graph. **23** (3), 399–405 (2004)
26. Thomas, D., Taniguchi, R.I.: Augmented blendshapes for real-time simultaneous 3D head modeling and facial motion capture. In: IEEE Conference on Computer Vision and Pattern Recognition, pp. 3299–3308 (2016)
27. Tulyakov, S., Vieriu, R.L., Semeniuta, S., et al.: Robust real-time extreme head pose estimation. In: IEEE International Conference on Pattern Recognition, pp. 2263–2268 (2014)
28. Wan, X., Jin, X.G.: Data-driven facial expression synthesis via Laplacian deformation. Multimedia Tools Appl. **58**(1), 109–123 (2012)

29. Wang, K.K., Zhang, G.F., Bao, H.J.: Robust 3D reconstruction with an RGB-D camera. IEEE Trans. Image Process. **23**(11), 4893–4906 (2014)
30. Waters, K.: A muscle model for animation three-dimensional facial expression. ACM SIGGRAPH **21**(4), 17–24 (1987)
31. Weise, T., Bouaziz, S., Li, H., et al.: Realtime performance based facial animation. ACM Trans. Graph. **30**(4), article 77 (2011)
32. Weng, Y.L., Cao, C., Hou, Q.M., et al.: Real-time facial animation on mobile devices. Graph. Models **76**(3), 172–179 (2014)
33. Zollhöfer, M., Martinek, M., Greiner, G., et al.: Automatic reconstruction of personalized avatars from 3D face scans. Comput. Animation Virtual Worlds **22**(2–3), 195–202 (2011)
34. Zollhöfer, M., Nießner, M., Izadi, S., et al.: Real-time non-rigid reconstruction using an RGB-D camera. ACM Trans. Graph. **33**(4), article 156 (2014)

The manufacturer's authorised representative in the EU is Springer Nature Customer Service Centre GmbH, Europaplatz 3, 69115 Heidelberg, Germany. If you have any concerns regarding our products, please contact ProductSafety@springernature.com

Printed and bound by CPI Group (UK) Ltd, Croydon, CR0 4YY
30/09/2026
02225330-0002